Laboratory Manual for
Physical Geology

McGraw-Hill Higher Education

A Division of The ***McGraw-Hill*** *Companies*

LABORATORY MANUAL FOR PHYSICAL GEOLOGY,
UPDATED TENTH EDITION

Published by McGraw-Hill, an imprint of The McGraw-Hill Companies, Inc., 1221 Avenue of the Americas, New York, NY 10020.

This book is printed on recycled, acid-free paper containing 10% postconsumer waste.

3 4 5 6 7 8 9 0 KGP/KGP 0 9 8 7 6 5 4 3 2 1

ISBN 0–07–366179–1

Vice president and editor-in-chief: *Kevin T. Kane*
Publisher: *JP Lenney*
Sponsoring editor: *Robert Smith*
Editorial assistant: *Jenni Lang*
Developmental editor: *Renee Russian*
Senior project manager: *Kay J. Brimeyer*
Associate media producer: *Judi David*
Production supervisor: *Sandy Ludovissy*
Coordinator of freelance design: *Rick D. Noel*
Cover design: *Annis Wei Leung*
Cover image: *FPG International © Otto Eberhard, Moraine Lake, Valley of the Ten Peaks, Banff NP, Alberta, Canada*
Senior photo research coordinator: *Lori Hancock*
Senior supplement coordinator: *David A. Welsh*
Compositor: *Shepherd, Inc.*
Typeface: *10/12 Times Roman*
Printer: *Quebecor Printing Book Group/Kingsport*

www.mhhe.com

Laboratory Manual for Physical Geology

Updated Tenth Edition

The Late
James H. Zumberge

Robert H. Rutford
University of Texas, Dallas

James L. Carter
University of Texas, Dallas

Boston Burr Ridge, IL Dubuque, IA Madison, WI New York San Francisco St. Louis
Bangkok Bogotá Caracas Lisbon London Madrid
Mexico City Milan New Delhi Seoul Singapore Sydney Taipei Toronto

Dedicated to the memory of

JAMES H. ZUMBERGE

whose dedication to his students

and commitment to the geosciences

live on

CONTENTS

PART 5
PLATE TECTONICS AND RELATED GEOLOGIC PHENOMENA 241

MATERIALS NEEDED BY STUDENTS USING THIS MANUAL

1. 10× hand lens.
2. Scale ("ruler") graduated in tenths of an inch.
3. Colored pencils (red, blue, and assorted other colors).
4. Felt tip pens (1/8″ × 1/4″ tip), three assorted colors.
5. Several medium to medium-soft pencils (2H or No. 2).
6. Small magnifying glass (optional) for map reading.
7. Six sheets 8 1/2″ × 11″ tracing paper.
8. Eraser (art gum or equivalent).
9. Inexpensive pencil sharpener.
10. Inexpensive compass, for drawing circles.
11. Dividers (optional), for measuring distances on maps.

PREFACE

The geological sciences continue to undergo remarkable changes. Those changes that have endured over time have been incorporated in each edition of this manual since the first edition was published in 1951. Although the subject matter has changed and expanded in scope, the number of laboratory sessions in a given academic quarter or semester has not increased. Because the time available in a quarter or semester cannot be expanded without disrupting the class schedule for the entire college or university, the problem of too much material for too little time poses a dilemma for authors, instructors and students.

On the assumption that the subject matter to be covered in any course is the prerogative of the instructor and not the authors, we have written a manual that contains more material than can be covered in a single laboratory course, thereby leaving the selection of individual exercises to the instructor. While we believe that the overall scope of this manual is in keeping with the general subject material covered in a beginning laboratory course, we think the instructors should determine the specific exercises that are in keeping with their own ideas of how to organize and present subject material.

In addition to the variety of laboratory exercises offered, we also provide background material with each exercise. To supplement this background material we have added a Glossary to this edition of the manual. By allowing students to review the important concepts and geologic terms they will encounter in the laboratory, we hope to enhance their chances for successful completion of the exercises. According to reviewers and users of past editions, the supplemental information provided is particularly useful in those instances where students do not routinely bring their textbooks to class or where students are not concurrently enrolled in the lecture course.

In the tenth edition we have added references to the World Wide Web in the form of "Web Connections." To assist students in the use of the Web, we have also included a new section called "Web Working 101." We recognize that not all students will have easy access to the Web, and so we have used the Connections as supplemental to the other background material provided for the exercises. Some instructors may wish to utilize materials from the Web Connections as part of the assigned laboratory exercises.

The tenth edition follows the same overall organization of the ninth edition. Past users will note that we now provide answer sheets for all of the exercises. Part 1, "Earth Materials," remains largely unchanged from the last edition. Many of our users tell us that they have developed their student collections to match the specimens used in this manual. They will see some minor, but no substantial, changes here. We have added, at the request of many users, tearout worksheets for mineral and rock identification.

Part 2, "Topographic Maps, Aerial Photographs, and Other Imagery from Remote Sensing," has been updated and expanded, especially the coverage of map coordinates, land divisions, and map projections.

The exercises in Part 3, "Geologic Interpretation of Topographic Maps, Aerial Photographs, and Earth Satellite Images" have had some refinements, but the content of the exercises remains unchanged.

Several of the figures in Part 4, "Structural Geology" have been revised for purposes of clarification, but again the content remains little changed from the ninth edition. Part 5, "Plate Tectonics and Related Geologic Phenomena" is unchanged.

We acknowledge with special thanks the graduate teaching assistants at The University of Texas at Dallas who have assisted with the revisions of this manual. Sincere thanks to Dr. Chris Parr for "Web Working 101."

We are especially grateful to Charles G. Sammis of the University of Southern California and Peter J. Wyllie of the California Institute of Technology for their help in preparing the materials on earthquakes and plate boundaries, respectively.

To those who reviewed this and past editions, we express our thanks and appreciation for their critical comments and suggestions for improvement. These include Dr. Mary Jo Richardson, Texas A&M University; Professor Vicki Harder, Texas A&M University; Professor Roseann J. Carlson, Tidewater Community College; Professor Harold Stowell, University of Alabama; Professor Ray Kenny, Arizona State University; Dr. Rudi H. Kiefer, University of North Carolina–Wilmington; Professor Anne Pasch, University of Alaska, Anchorage; Dr. Thomas E. Hendrix, Grand Valley State University; and Dr. Phillip R. Kemmerly, Austin Peay State University.

As authors we accept the full responsibility for any inadvertent errors that have crept into these pages, and we welcome comments from users if they discover such errors. We also hope that users will make suggestions to us that will assist us in the continued improvement of this manual in the future.

Finally, we extend our gratitude to the professional men and women of McGraw-Hill for their design of the format and expert help in transforming our manuscript into a final product.

ROBERT H. RUTFORD
University of Texas at Dallas

JAMES L. CARTER
University of Texas at Dallas

WEB WORKING 101

INTRODUCTION

The *World Wide Web* (WWW or W^3) is one of the most exciting new developments in education in decades. With it, you can extract an ever-growing wealth of information and no small quantity of nonsense as well! This is because the Web permits unreviewed and uncensored authorship to all who seek to publish. In that sense, it is unlike your textbook, which has been edited and reviewed by many to ensure both accuracy and efficacy. But also unlike your textbook, the Web is vast, encyclopedic, and unbound by the usual production schedules of publication; so information can be found on the Web that is quite current.

It is ever-changing, not only in content but also in style and access. The Web you grow used to today will certainly not be the same entity your children surf in the future. So using the Web implies a willingness not only to learn new facts and theories but also new methods of finding and absorbing them as well.

This document seeks to help you to use the Web fearlessly and with fun as you supplement your coursework with greater breadth and depth in areas of interest to you. It also drops hints about why various aspects of the Web have developed as they have. And while this document is in the nature of a tutorial, you should be confident about the fact that the Web is the most heavily self-documented entity in the known Universe! So when you come to the end of this document and find yourself hungry for more, the Web will sate you.

BASICS

A **Web Page** differs from your textbook page in interesting ways. You can't "turn the page" of your computer screen, so advancing through its content requires a different metaphor. The current metaphor is taken from publication more ancient than books, namely scrolls. As an ancient Egyptian, you would have unrolled your papyrus as you read; as a modern Computati, you instead drag a scroll control knob down the vertical scroll bar (screen right) by holding your mouse's left button down while on the knob. If the page is wider than your screen, you'll see a horizontal scroll bar (screen bottom) and its knob to drag right to reveal hidden page content; technically that's "panning" (as with a movie camera) rather than "scrolling." But if you've done any word processing, you've already mastered both skills.

Likewise, if you've run any applications in a windowing environment, you will recognize the **title bar** as the shorthand reference for what the application (or in this case, the web page) is about. It takes on a greater importance on the Web since keywords that appear are often those given more significance in keyword searching.

What may be new to you, if you're just beginning your web working, are the ***hypermedia links.*** These underscored phrases are, perhaps, the most basic feature of the Web, since they serve as the portals to jump between documents all over the world! When you encounter a link you wish to explore, your mouse pointer, resting on the link text, changes its shape to make you realize that here is a jumping-off point. Clicking your mouse then instructs your Browser to load the referenced web page, and you're off!

But text phrases are not the only link objects you may find. Images and, more commonly, **parts of images** can serve as linking portals. Again, the change in your mouse pointer as you pass it over these objects will give you a clue that they *are* links. The images themselves may suggest where you'll go, but there is another clue. A commentary line on your Browser (screen bottom) will display the address of the *link target* while your mouse pointer rests on the link. This should not be confused with the address of the *current* web page, which will usually be displayed at the top of the Browser's screen.

Both such addresses will have a special form known (a bit grandiosely) as a **Universal Resource Locator** or **URL.** The usual form you will see looks like **http://www.institution.type/directory/webpage.html** which will be the form that your textbook's author will give to refer you to interesting sites that help illuminate the text's discussion or take you deeper into the subject. It is worth understanding the construction of such addresses because then you can make up likely addresses to browse on your own!

BROWSING

The URL tells your Browser to command the computer having the data you desire to send it to you. That target computer will be in some institution on the **www,** and the institution will be of some **type,** such as educational or commercial, for example. If the institution is not in the United States, there will be a country code, such as **ca** for Canada, as well. The computer that is serving you the data you wish, called the Server for want of imagination, has the information in its mass storage, organized in some **directory** structure for convenience. And if, as is likely, it is another web page, it will be made up from the language of web pages called HyperText Markup Language, or HTML for short. It will be a file within the directory with a name and designation of **html** or perhaps **htm.** While it's simple to learn, you needn't know HTML in order to recognize that that part of the filename means that it *is* a web page.

When you give your Browser application a sample address from your text, perhaps, you are starting on a journey to places that will reference other places to send you on, to still more places, almost without end. Each web page you visit will normally have multiple links to similarly linked pages, and the opportunities to get lost abound! So almost as important as starting and continuing your journey is **how to get back home.** Your Browser will remember each journey you begin and the links between all of the web pages that you visit. You can take advantage of that in two ways.

First, you can make the Browser **Go Back** a link at a time with a new menu item by such a name. This is useful not only to get home but also to browse further unexplored links in previous web pages.

Second, from another menu item you can see a listing of the last several web pages and jump to any of them by clicking on their titles in the list. In either case, on backing up, you'll discover that a menu item like **Go Forward** becomes active to permit you to shuttle up and down the chain of links you've been following.

The ancient Chinese proverb is still right: *A journey of 1,000 miles begins with a single step.* So you must know how to take that first step. And the place to do that is in your Browser's window, where the *current* web page address is displayed. You must replace that address with the one you wish to follow such as those from among your author's suggestions. Highlighting that address line with your mouse makes it vulnerable to change just as it does with word processing text. Typing the chosen address and hitting the Enter key does the trick. The only other trick is to type the address both completely and correctly. Do not forget the prefix **http://** and *do* use ***capitalization*** only where it is present in the address. Most servers are sensitive to the case!

That **http://** stands for **HyperText Transport Protocol,** but it isn't the only way to scamper around the Web. The Web evolved from earlier forms of the internet designed to transport other kinds of documents and commands. You will, no doubt, on your journey see other such protocols like the ubiquitous **ftp://** or File Transfer Protocol, which is very useful in retrieving files destined for use by applications other than your Browser. For example, your instructor might create a word processor or presentation document for you to read or display with your own applications; those will likely come to you via **FTP.**

One of the most interesting features of the Web is the eagerness with which vendors of document creation software offer viewer applications for these documents *free* to all who wish them! It's a bit as if the manufacturers of TV equipment were to give away TV sets but sell TV cameras. The giveaway helps create a demand for programming, which can only be met by using the cameras to create the content. But TV signal standards change infrequently, so the manufacturers can sell both camera *and* receiver. Web document types are in unbelievably rapid flux; so to sell content creation software, vendors must ensure a demand with this free giveaway of viewers. Web pages offering different document types will have links to these free resources so that you can view their special contents.

Often you browse this extra-HTML content with **"helpers"** to your Browser. The rapid evolution of browsers usually means incorporation of popular document viewers, but others still arise that offer to embed their viewer to make it available as if it were native to your Browser. The alternative is to open such viewers as windows outside your Browser, requiring you to navigate between applications as well as between web pages.

While HTML and other kinds of documents can come to you as you follow the trail of web page links in your browsing mode, you can almost always find the content you wish *even* if you do not know where the URL resides. This is because the Web is already too extensive and too rapidly growing for even your textbook's author to know it all. But that isn't a daunting prospect since one of the first features to arise on the internet was the ability to *hunt* for information of interest based upon key words, phrases, or even file names! So to take the greatest advantage of the Web, you must learn to search the Web.

This is possible because many servers have automated search engines, called robots, which mindlessly and tirelessly scan the growing list of web pages for new entries, treating them as they have ancient, even now nonexistent (!), pages by cataloguing their words and phrases in giant databases for rapid retrieval of their addresses given the stimulus of your request. Other sites take the more thoughtful but less thorough approach of reviewing worthwhile web pages for inclusion in categories through which you may browse. In either case, search strategies are most definitely your friends!

SEARCHING

Hunting for information on the Web can be very similar to hunting in your school library's electronic catalog. You can often use not only key words and phrases but also combinations of such keys with logical operations like **AND, OR, NOT,** etc. and fuzzy logical operations like **NEAR.** These are elements of what you may already know to be a Boolean Search. An example of such a search might be **wonderland AND NOT alice,** which would exclude web pages that reference *"Alice in Wonderland"* and find others like *"Winter Wonderland."* You should resist the temptation to ask for **wonderland BUT NOT alice** since **BUT** isn't recognized as a Boolean operation.

Properly formed searches are as much an art as a science, since the results that come back to you can be analyzed to suggest more fruitful strategies with greater or lesser exclusions. Often the searching sites make extensive help screens available to tutor you in taking the most advantage of that site's particular implementation of searching. If the site offers a search on exact phrases, it will likely return the most focused results. Exact phrases are often recognized by the searcher as those enclosed in quotes; so the same request above when enclosed in quotes, **"wonderland AND NOT alice,"** is *not* a Boolean search because **AND NOT** is part of the phrase you appear to be seeking. Since it is highly unlikely that any of the millions of web pages contain that exact quote, no web page will be found. (So don't expect Boolean operations to work *inside* quoted phrases.)

Searching is so fundamental to Web Working that web pages that reference web pages where searching is done are a rich source of help. One very helpful one is at
http://www.scout.cs.wisc.edu/scout/toolkit/searching/index.html
(which is served from the Computer Science Department at The University of Wisconsin in cooperation with the National Science Foundation and two commercial communications enterprises). In addition, some web pages such as
http://www.metacrawler.com/
pass your search request to half a dozen or more other search engines, culling out duplicate responses and presenting you, as do all successful search results, with links to the discovered web pages. As you explore such suggestions, remember to **Go Back** to the search results page to try additional links. Be aware that search results often find hundreds of thousands of "hits" but display only the first 10 or 20 of them; so look for links to the next 10 or 20 so as not to miss an important lead.

Your Browser assists you even further as you explore leads for it and permits you to save **Bookmarks** of interesting web page addresses at your command. As your searching and browsing becomes more complex, if not obsessional, you should organize your Bookmarks in categories if your Browser permits it or in separate Bookmark files if it does not. Clicking on a Bookmark address will, of course, transport you to that address. That's its purpose.

With so many search engines to choose from, you should browse several to find the one that works the way you like to think. For example, if your searches take you to web pages written in foreign languages, the Alta Vista search engine,
http://www.altavista.digital.com/
would be advantageous, since it has a rudimentary translation program to help you understand the content in several languages.

In addition to key word search databases, you should explore web pages that offer subject catalogs and guides since they offer encyclopedic categories to browse and even commentary on the content to be found on many links. The previously mentioned Scout Toolkit is a good starting place to seek such references.

For a very large collection of search engines and databases, see the site maintained by William Cross at
http://www.albany.net/allinone/
and to better understand how to evaluate search engines, try the site maintained by Library & Media Services at the University of Detroit Mercy, which is located at
http://www.udmercy.edu/htmls/Academics/library/search.

Citing Sites

While you will undoubtedly browse and search the Web for the sheer joy of it, there'll come a time when your research is focused on a term paper or other document derived in part from the results of Web Working. In order not to misrepresent the works of others, you must cite the source of all material you use in your report. While each discipline has its preferred format for citations, they all involve identification of authorship, title, publication, and its date. Given the rapid appearance and disappearance (or evolution) of web pages on the internet, you should also note the date you visited the site.

The author of the page could be either its creator or someone identified on the page as the maintainer of the information. The title is best given as that on the title line of the Browser's window. The publication is given as its *valid* URL. And its date might be either a creation or a revision date, usually given at the end of the web page.

If your report is itself prepared as a document file, remember to separate the URL from any punctuation so that it stands alone as a complete address including the proper prefix, e.g., **http://** if it's a web page. That isolation will permit contemporary document readers to identify the URL as an internet reference and turn it automatically into a link to the referenced page. Your instructor will be able to jump to it and will appreciate your thoughtfulness!

Caveats

Caveat, of course, comes from the famous Latin phrase, "*Caveat Emptor,*" or "*Let the buyer beware.*" And there are plenty of warnings to go around about the Web! We've all heard what a Wild West place it is, abounding in pornography, hate literature, and the like. But cautions against surfing in nasty neighborhoods are not what we have in mind here. Instead we want to alert you to more mundane problems with the Web and hope to overcome them.

First, the Web is a very busy place. Traffic jams on it are bound to frustrate. So when your surfing slows to a crawl, what can you do? You can recognize that pictures are worth even more than a thousand words and tell your Browser that, temporarily, you'd like to not waste time on glitzy photos and op-art logos until you find a site whose information you'd like to experience as a whole. You do that by locating and exercising your **Preferences** in your Browser. Thereafter, until you enable graphics again, the web pages will fill quickly with text and display an icon where images would have been. Often within the borders of what would have been pictures will be brief text descriptions of what you're missing instead. When you find a web page whose graphics are worth the wait, you can change the Preference back and tell your Browser to **Reload** the page.

The only danger you run in rejecting images occurs when a web site uses **image maps** for navigation. These are pictures within which different areas are links to different web pages, the subimages suggestive of where you'll go when you click on them. Without the image, an image map is worthless. So if your mouse cursor identifies different link targets (in the Browser's comment line) as it crosses an unloaded picture area, you'd best enable graphics and **Reload** that page.

But even this fix presumes that you can connect to a target site's server. When that connection seems to be taking too long (in your judgment), is there anything you can do? Often a fresh request for a web page will bump you to a higher priority than one that appears to be ignored. So you have to terminate the link request and click on the link again. Termination is easy once you've found your Browser's **Stop** button (a wonderfully human innovation). One of two things will happen when you **Stop.** If the target site has not been contacted or has not responded to initial contact, you'll be left at the calling page, ready to click on the link again. But if the target has been dribbling its information to you, the **Stop** will interrupt the transfer and the Browser will display as much of the target page as it has received. So if you see an incomplete receipt, simply **Reload** the target to try to obtain higher priority. Of course, it's good to remember that a busy site may become less busy in some off hours; so what time *is* it at that site's institution anyway?

Popular information is often **"mirrored"** by having it available on many servers at many institutions. When you request such information, you'll be asked to choose the site nearest you. If another link to the same data is pointed at an institution elsewhere on the globe where it is now the middle of the night, you might actually get it faster from the bored computer there! That is a stunning reminder that you live in a small world that is getting smaller as the Web expands to cover it.

By far, the most frightening part of Web Working is the realization that if you can connect to control computers halfway around the planet, they can connect to control yours! When you plug into the web, it's a two-way street. Your **Internet Service Provider (ISP)** will protect you by changing the identity of your connection even while you are using it; when that happens, you don't notice a thing, but a hacker who thought he had you in his sights, suddenly finds you gone. While that *is* a comfort, it means that legitimate sites to which you connect cannot use your connection address to identify you and keep track of what you want; you keep disappearing on them too.

They counter by offering you **Cookies.** Not oatmeal or chocolate chip, these cookies are codes that get stored in your Browser to be passed back to the web site for continuity of identification. Another **Preference** in your Browser will tell it to notify you of these offers so that you can accept some and reject others (where your identity need not be an issue). But if a cookie notice is not enabled, these little recognition codes will fly back and forth automatically. How paranoid you want to be over such identification is up to you.

OK, what happens if you encounter a rogue web site intent on causing mischief in your computer? What weapons can it bring to bear? If the site offers programs to "download" to your computer, those could carry computer viruses; so practice safe sectors and don't run them. No problem.

But can that rogue do damage *before* you can act? The process of receiving HTML codes is completely without risk as long as you reject program offers. However, the current trend in web communications stems from a realization that useful data can often be obtained from programs smaller than the answers they produce! In which case, it is far faster to have your Browser run such small programs locally than it would be to transfer their voluminous answers. The language in which these are written is **Java,** and it is possible to write what amount to Java viruses. So it may be prudent to disable Java in your Browser until you arrive at a *trusted* site that needs it to help you. Such mischief opportunities will certainly diminish as they are found and written out of future versions of Java. (Internet Explorer's **Active-X** has more serious security problems, but it cannot be disabled; so IE users must learn to live with potential hazards.)

Part One
Earth Materials

Background

The materials that make up the crust of the earth fall into two broad categories: minerals and rocks. Minerals are elements or chemical compounds that are formed by a number of natural processes. Rocks are aggregates of minerals or organic substances that occur in many different architectural forms over the face of the earth, and they contain a significant part of the geologic history of the region where they occur. To identify them and understand their history, it is necessary to be able to identify the minerals that make up the rocks.

The first goal of Part One is to introduce beginning students of geology to the identification of minerals and rocks through the use of simplified identification methods and classification schemes. Students will be provided with samples of minerals and rocks in the laboratory. These samples are called *hand specimens.* Ordinarily their study does not require a microscope or any means of magnification because the naked or corrected eye is sufficient to perceive their diagnostic characteristics. A feature of a mineral or rock that can be distinguished without the aid of magnification is said to be *macroscopic* (also *megascopic*) in size. Conversely, a feature that can be identified only with the aid of magnifiers is said to be *microscopic* in size. The exercises that deal with the identification and classification of minerals and rocks in Part One are based only on macroscopic features.

The second goal of this part is to acquaint the beginning student with the modes of formation and occurrence of rocks and their relative age relationships. Many beginning students using this manual will participate in organized field trips to observe firsthand how rocks occur in nature. To prepare students for this field experience, some basic geological principles on the mode of occurrence of rocks will be introduced in the form of simple geologic diagrams. The concept of geologic time will be explored, and the relative age relationships of rock masses will be examined according to some basic geologic principles and assumptions.

Even if organized field trips are not a required part of your geology course, an understanding of these geologic diagrams will enhance the appreciation of rock strata and other geologic phenomena by your students as they encounter them in their travels.

MINERALS

DEFINITION

A mineral is a naturally occurring, crystalline, inorganic, homogeneous solid with a chemical composition that is either fixed or varies within certain fixed limits, and a characteristic internal structure manifested in its exterior form and physical properties.

MINERAL IDENTIFICATION

Common minerals are identified or recognized by testing them for general or specific physical properties. For example, the common substance table salt is actually a mineral composed of sodium chloride (NaCl) and bears the mineral name halite. The taste of halite is distinctive and is sufficient for identifying and distinguishing it from other substances such as sugar (not a mineral) that have a similar appearance. Chemical composition alone is not sufficient to identify minerals. For example, the mineral graphite and the mineral diamond are both composed of a single element, carbon (C), but their physical properties are very different.

The taste test applied to halite is restrictive because it is the only mineral that can be so identified. Other minerals may have a specific taste, but it is different from that of halite. Other common minerals can be tested by visual inspection for the physical properties of crystal form, cleavage, or color or by using simple tools such as a knife blade or glass plate to test for the physical property of hardness.

The first step in learning how to identify common minerals is to become acquainted with the various physical properties that individually or collectively characterize a mineral specimen.

PROPERTIES OF MINERALS

The physical properties of minerals are those that can be observed generally in all minerals. They include such common features as luster, color, hardness, cleavage, streak, and specific gravity. *Special properties* are those that are found in only a few minerals. These include magnetism, double refraction, taste, odor, feel, and chemical reaction with acid. In your work in the laboratory, use the hand specimens sparingly when applying tests for the various properties.

GENERAL PHYSICAL PROPERTIES

Luster

The appearance of a fresh mineral surface in reflected light is its luster. A mineral that looks like a metal is said to have a *metallic luster.* Minerals that are *nonmetallic* are described by one of the following adjectives: *vitreous* (having the luster of glass); *resinous* (having the luster of resin); *pearly; silky; dull* or *earthy* (not bright or shiny).

Your laboratory instructor will display examples of minerals that possess these various lusters.

Color

The color of a mineral is determined by examining a fresh surface in reflected light. Color and luster are not the same. Some minerals are clear and transparent, thus colorless. The color or lack of color may be diagnostic in some minerals, but in others the color varies due either to a slight difference in chemical composition or to small amounts of impurities within the mineral (fig. 1.1).

Hardness

The hardness of a mineral is its resistance to abrasion. Hardness can be determined either by trying to scratch a mineral of unknown hardness with a substance of known hardness, or by using the unknown mineral to scratch a substance of known hardness. Hardness is measured on a relative scale called the *Mohs scale of hardness,* which con-

TABLE 1.1 *Mineral Hardness According to the Mohs Scale (A) and Some Common Materials (B)*

HARDNESS	A	B
1	Talc	
2	Gypsum	
2.5		Fingernail
3	Calcite	
3.5		Copper penny
4	Fluorite	
5	Apatite	
5–5.5		Knife blade
5.5		Glass plate
6	Orthoclase	
6.5		Steel file
7	Quartz	
8	Topaz	
9	Corundum	
10	Diamond	

sists of ten common minerals arranged in order of their increasing hardness (table 1.1). In the laboratory, convenient materials other than these ten specific minerals may be used for hardness determination.

In this manual, a mineral that scratches glass will be considered "hard," and one that does not scratch glass will be "soft." In making hardness tests on a glass plate, do not hold the glass in your hands; keep it firmly on the tabletop. If you think that you have made a scratch on the glass, try to rub the scratch off. What appears to be a scratch may be only some of the mineral that has rubbed off on the glass.

Cleavage

Cleavage is the tendency of a mineral to break along definite planes of weakness that exist in the internal (atomic) structure of the mineral. Cleavage planes are related to the crystal system of the mineral and are always parallel to crystal faces or possible crystal faces. Cleavage may be conspicuous and is a characteristic physical property that is useful in mineral identification. It is almost impossible to break some minerals in such a way that cleavage planes do not develop. An example is calcite, with its rhombohedral cleavage.

Perfect cleavage describes cleavage planes that are very smooth and flat and that reflect light much like a mirror. Other descriptors such as *good, fair,* and *poor* are used to describe cleavage surfaces that are less well defined. Some minerals exhibit excellent crystal faces but have no cleavage; quartz is such a mineral.

The cleavage planes of some minerals such as calcite, muscovite (fig. 1.2), halite, and fluorite are so well developed that they are easily detected. In others, the cleavage surfaces may be so discontinuous as to escape detection by casual inspection. Before deciding that a mineral has no cleavage, turn it around in a strong light and observe whether there is some position in which the surface of the specimen "lights up," that is, reflects the light as if it were the reflecting surface of a dull mirror. If so, the mineral has cleavage, but the cleavage surface consists of several discontinuous parallel planes minutely separated, and rather than perfect cleavage it has good, fair, or poor cleavage. As will be noted in the discussion of crystal form, it is important to differentiate between cleavage planes and crystal faces; the actual breaking of a mineral crystal may be useful in making this differentiation.

FIGURE 1.1

The specimens in this photograph are all quartz. The difference in colors is due to various impurities. Clockwise from left: smoky quartz, quartz crystal, rose quartz, citrine quartz, amethyst quartz.

FIGURE 1.2

Muscovite is a mineral with excellent basal cleavage.

In assigning the number of cleavage planes to a specimen, do not make the mistake of calling two parallel planes bounding the opposite sides of a specimen two cleavage planes. In this case, the specimen has two cleavage surfaces but only one plane of cleavage (i.e., one direction of cleavage). For example, halite has cubic cleavage, thus six sides, but only three planes of cleavage, because the six sides are made up of three parallel pairs of cleavage planes.

The angle at which two cleavage planes intersect is diagnostic. This angle can be determined by inspection. In most cases you will need to know whether the angle is 90 degrees, almost 90 degrees, or more or less than 90 degrees. The cleavage relationships that you will encounter during the course of your study of common minerals are tabulated for convenience in figure 1.3.

Some minerals exhibit the characteristic of *parting,* sometimes called false cleavage. Parting occurs along planes of weakness in the mineral, but usually the planes are more widely separated and often are due to twinning deformation or inclusions. Parting is not present in all specimens of a given mineral.

Fracture generally refers to breakage that forms a surface with no relationship to the internal structure of the mineral; that is, the break occurs in a direction other than a cleavage plane. Quartz is characterized by *conchoidal* fracture (the fracture surfaces are smooth and exhibit fine concentric ridges); asbestos by *fibrous* fracture. Other terms often used include *hackly, uneven* (rough), *even* (smooth), and *earthy* (dull but smooth fracture surfaces, common in soft mineral aggregates such as kaolinite).

Streak

The color of a mineral's powder is its streak. The streak is determined by rubbing the hand specimen on a piece of unglazed porcelain (*streak plate*). Some minerals have a streak that is the same as the color of the hand specimen; others have a streak that differs in color from the hand specimen. The streak of minerals with a metallic luster is especially diagnostic.

Specific Gravity

The specific gravity (G) of a mineral is a number that represents the ratio of the mineral's weight to the weight of an equal volume of water. In contrast to density, defined as weight per unit volume, specific gravity is a dimensionless number. The higher the specific gravity, the greater the density of a mineral.

For purposes of determining specific gravity of the minerals in the laboratory, it is sufficient to utilize a simple

Number of Cleavage Planes	Remarks	Examples
1	**Usually called *basal cleavage.*** Examples: muscovite and biotite.	
2	**Two at 90 degrees.** Examples: feldspar and pyroxene (augite) have cleavage surfaces that intersect at close to 90 degrees.	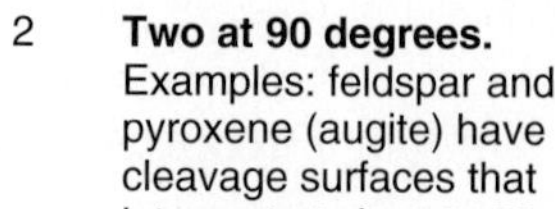 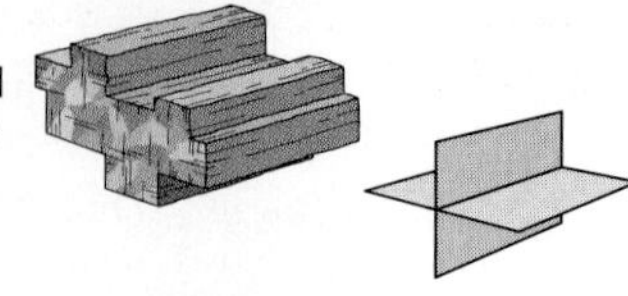
2	**Two *not* at 90 degrees.** Example: amphibole (hornblende) has cleavage surfaces that intersect at angles of about 60 and 120 degrees.	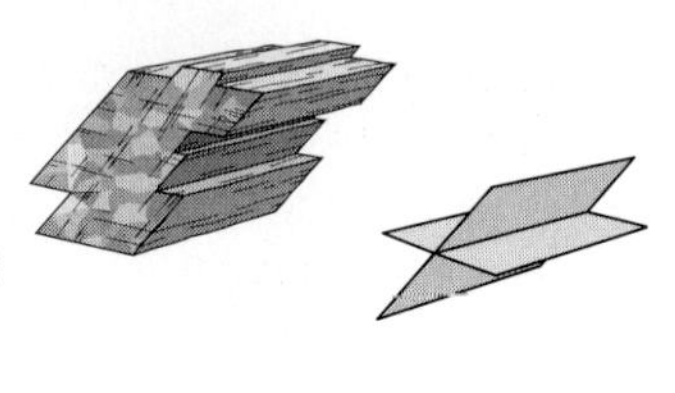
3	**Three at 90 degrees. Minerals with three planes of cleavage that intersect at 90 degrees are said to have *cubic cleavage.*** Example: halite and galena.	
3	**Three *not* at 90 degrees. A mineral that breaks into a six-sided prism, with each side having the shape of a parallellogram, has *rhombic cleavage.*** Example: calcite.	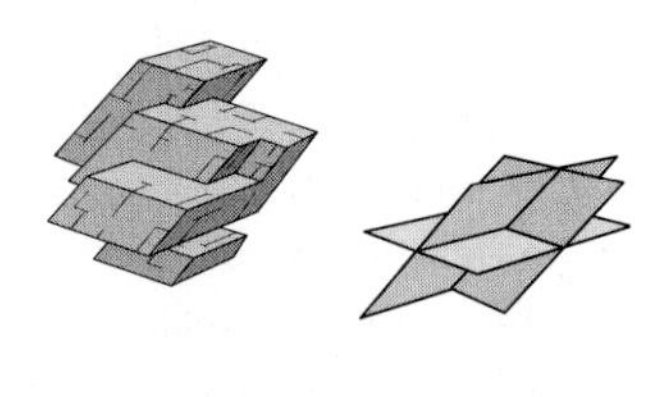
4	**Four sets of cleavage surfaces in the form of an octahedron produce *octahedral cleavage.*** Example: fluorite.	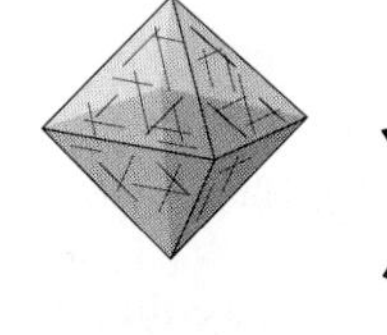
6	**Complex geometric forms.** Example: sphalerite.	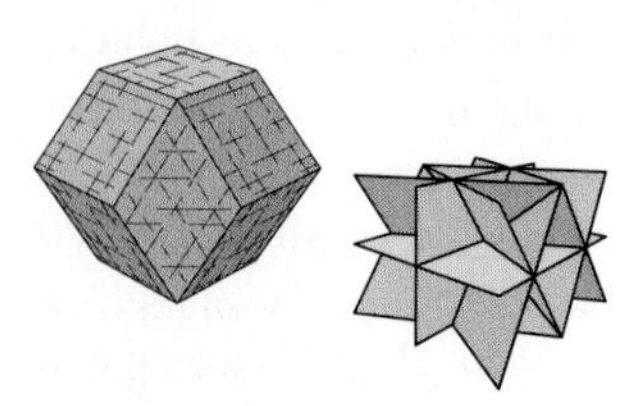

FIGURE 1.3

Descriptive notes on cleavage planes.

(Adapted and reprinted with permission from R. D. Dallmeyer, Physical Geology Laboratory Manual. Copyright © 1978 by Kendall/Hunt Publishing Company.)

"heft" test by lifting the hand specimens. Compare the heft of two specimens of about the same size but of contrasting specific gravity. For example, heft a specimen of graphite (G = 2.2) in one hand while hefting a similarly sized specimen of galena (G = 7.6) in the other. *Take care to compare only similar-sized samples.* This allows you to determine the relative specific gravity of minerals. Minerals such as graphite (G = 2.2) and gypsum (G = 2.3) are relatively light when hefted. Quartz (G = 2.6) and calcite (G = 2.7) are of average heft, whereas corundum (G = 4.0), magnetite (G = 5.2), and galena (G = 7.6) are heavy when hefted.

Diaphaneity

The ability of a thin slice of a mineral to transmit light is its diaphaneity. If a mineral transmits light freely so that an object viewed through it is clearly outlined, the mineral is said to be *transparent.* If light passes through the mineral but the object viewed is not clearly outlined, the mineral is *translucent.* Some minerals are transparent in thin slices and translucent in thicker sections. If a mineral allows no light to pass through it, even in the thinnest slices, it is said to be *opaque.*

Tenacity

This property is an index of a mineral's resistance to being broken or bent. It is not to be confused with hardness. Some of the terms used to describe tenacity are:

Brittle—The mineral shatters when struck with a hammer or dropped on a hard surface.

Elastic—The mineral bends without breaking and returns to the original shape when stress is released.

Flexible—The mineral bends without breaking but does not return to its original shape when the stress is released.

Crystal Form

A crystal is a solid bounded by smooth surfaces (crystal faces) that reflect the internal (atomic) structure of the mineral. *Crystal form* refers to the assemblage of faces that constitute the exterior surface of the crystal. *Crystal symmetry* is the geometric relationship between the faces.

Seven crystal systems are recognized by crystallographers, and all crystalline substances crystallize in one of the seven crystal systems (fig. 1.4). Some common substances, such as glass, are often described as crystalline, but in reality they are *amorphous*—they have solidified with no fixed or regular internal atomic structure.

The same mineral always shows the same angular relations between crystal faces, a relationship known as the *law of constancy of interfacial angles.* The symmetric relationship of crystal faces, related to the constancy of interfacial angles, is the basis for the recognition of the crystal systems.

Symmetry in a crystal is determined by completing a few geometric operations. For example, a cube has six faces, each at right angles to the adjacent faces. A planar surface that divides the cube into portions such that the faces on one side of the plane are mirror images of the faces on the other side of the plane is called a *plane of symmetry.* A cube has nine such planes of symmetry. In the same way, imagine a line (axis) connecting the center of one face on a cube with the center of the face opposite it. Rotation of the cube about this axis will show that during a complete rotation a crystal face identical with the first face observed will appear in the same position four times. This is a *fourfold axis of symmetry.* Rotation of the cube around an axis connecting opposite corners will show that three times during a complete rotation an identical face appears, thus a *threefold axis of symmetry.*

The seven crystal systems can be recognized by the symmetry they display. Figure 1.4 summarizes the basic elements of the symmetry for each system and shows some examples of the *crystal habit* (the crystal form commonly taken by a given mineral) of some minerals you may see in the laboratory or a museum.

Perfect crystals in nature are the exception rather than the rule. They usually form under special conditions where there is open space for them to grow during crystallization. Crystals more commonly are small and distorted. Nevertheless, the internal arrangement of the atoms is fixed, although the external form is not perfectly developed.

Many of the hand specimens you see in the laboratory will be made up of many minute crystals so that few crystal faces, or none, can be seen, and the specimen will appear granular. Other hand specimens may be fragments of larger crystals so that only one or two imperfect crystal faces can be recognized. Although perfect crystals are rare, most student laboratory collections contain some reasonably good crystals of quartz, calcite, gypsum, fluorite, and pyrite.

A word of caution: Cleavage fragments of minerals such as halite, calcite, and gypsum are often mistaken for crystals. This is because their cleavage fragments have the same geometric form as the crystal.

Two or more crystals of some minerals may be grown together in such a way that the individual parts are related through their internal structures. The external form that results is manifested in a *twinned crystal.* Some twins appear to have grown side by side (plagioclase), some are reversed or are mirror images (calcite), and others appear to have penetrated one another (fluorite, orthoclase, staurolite). Recognition of twinned crystals may be useful in mineral identification.

CRYSTAL SYSTEM	CHARACTERISTICS	EXAMPLES*
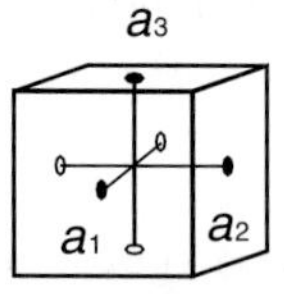 a_3, a_1, a_2 CUBIC (ISOMETRIC)	Three mutually perpendicular axes, all of the same length ($a_1 = a_2 = a_3$). Fourfold axis of symmetry around a_1, a_2, and a_3.	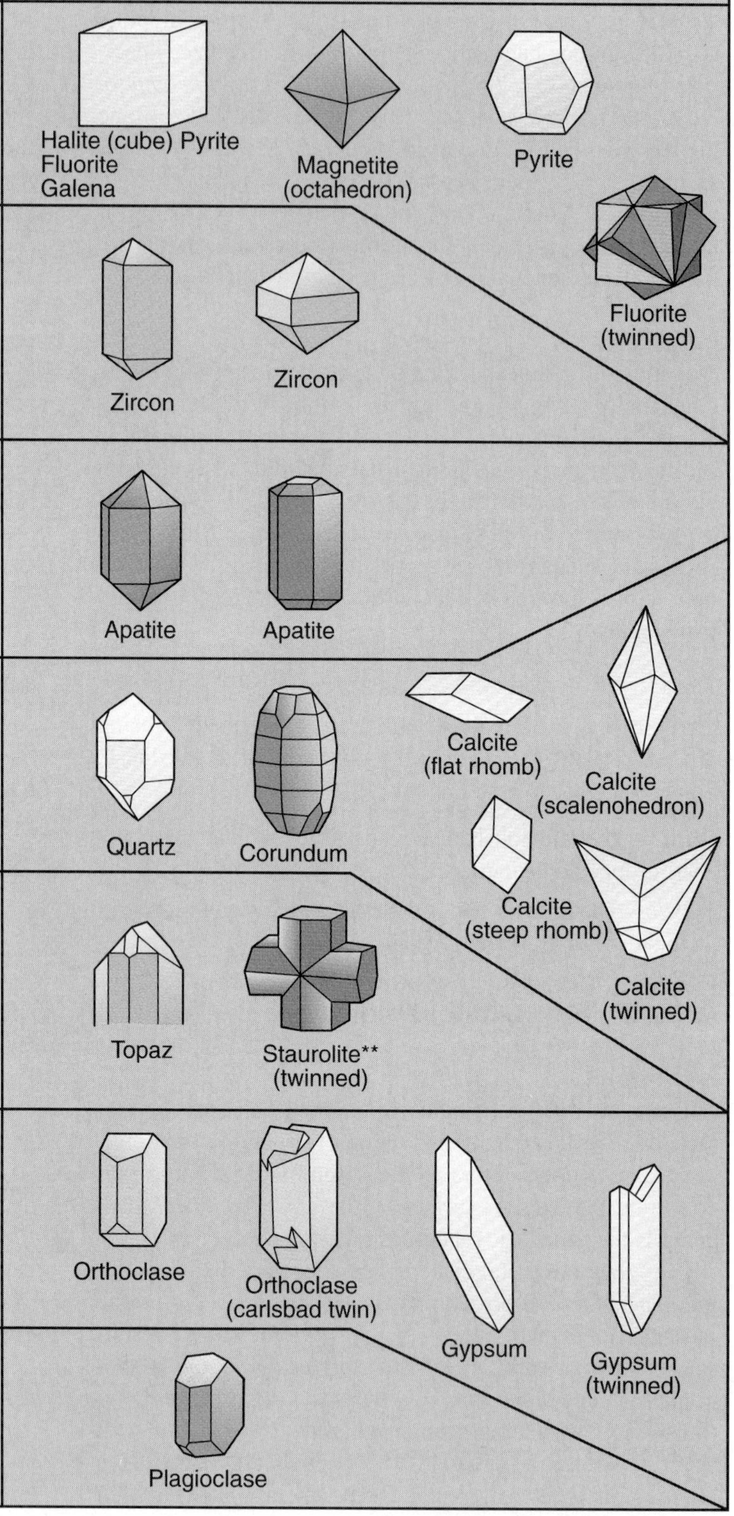 Halite (cube) Pyrite Fluorite Galena; Magnetite (octahedron); Pyrite; Fluorite (twinned)
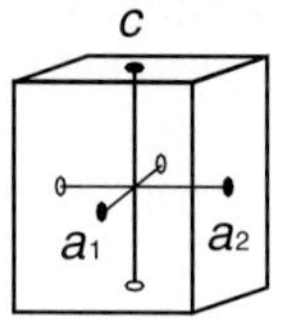 c, a_1, a_2 TETRAGONAL	Three mutually perpendicular axes, two of the same length ($a_1 = a_2$) and a third (c) of a length not equal to the other two. Fourfold axis of symmetry around c.	Zircon; Zircon
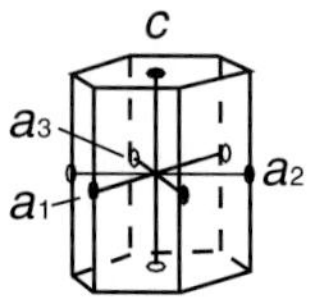 c, a_3, a_1, a_2 HEXAGONAL	Three horizontal axes of the same length ($a_1 = a_2 = a_3$) and intersecting at 120°. The fourth axis (c) is perpendicular to the other three. Sixfold axis of symmetry around c.	Apatite; Apatite
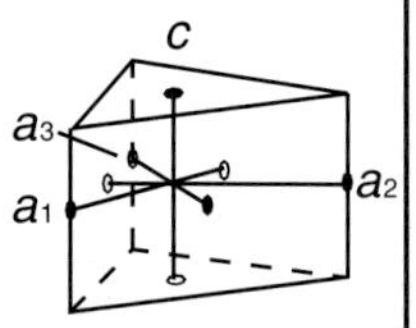 c, a_3, a_1, a_2 TRIGONAL	Three horizontal axes of the same length ($a_1 = a_2 = a_3$) and intersecting at 120°. The fourth axis (c) is perpendicular to the other three. Threefold axis of symmetry around c.	Quartz; Corundum; Calcite (flat rhomb); Calcite (scalenohedron); Calcite (steep rhomb); Calcite (twinned)
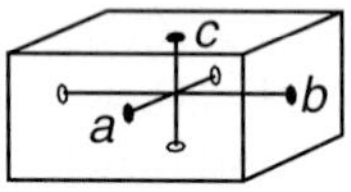 c, a, b ORTHORHOMBIC	Three mutually perpendicular axes of different length. ($a \neq b \neq c$). Twofold axis of symmetry around a, b, and c.	Topaz; Staurolite** (twinned)
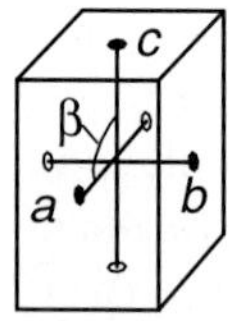 c, β, a, b MONOCLINIC	Two mutually perpendicular axes (b and c) of any length. A third axis (a) at an oblique angle (β) to the plane of the other two. Twofold axis of symmetry around b.	Orthoclase; Orthoclase (carlsbad twin); Gypsum; Gypsum (twinned)
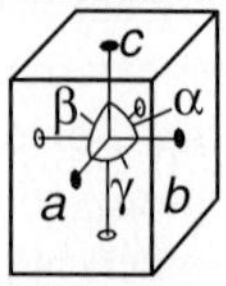 c, β, α, a, γ, b TRICLINIC	Three axes at oblique angles (α, β, and γ), all of unequal length. No rotational symmetry.	Plagioclase

FIGURE 1.4

Characteristics of the seven crystal systems and some examples.

*(Copyright © 1974 McGraw-Hill Book Co. (UK) Ltd. From Cox, Price & Harte: An Introduction to the Practical Study of Crystals, Minerals and Rocks, Revised Edition. Reproduced by permission. Colors have been added to the original and are not accurate. They are shown for illustrative purposes only.) *Most laboratory collections of minerals for individual student use do not include crystals of these minerals. The collection may, however, contain incomplete single crystals, fragments of single crystals, or aggregates of crystals of one or more minerals. The best examples of these and other crystals may be seen on display in most mineralogical museums. **Staurolite is actually monoclinic but is also classified as pseudo-orthorhombic. Pseudo-orthorhombic means that staurolite appears to be orthorhombic because the angle β in the monoclinic system (see left-hand column under monoclinic) is so close to 90 degrees that in hand specimens it is not possible to discern that the angle β for staurolite is actually 89 degrees, 57 minutes.*

Special Properties

Magnetism

The test for magnetism requires the use of a common magnet or magnetized knife blade. Usually, magnetite is the only mineral in your collection that will be attracted by a magnet.

Double Refraction

If an object appears to be double when viewed through a transparent mineral, the mineral is said to have double refraction. Calcite is the best common example (see fig. 1.13).

Taste

The saline taste of halite is an easy means of identifying the mineral. Few minerals are soluble enough to possess this property. (For obvious sanitary reasons, do not use the taste test on your laboratory hand specimens.)

Odor

Some minerals give off a characteristic odor when damp. Exhaling on the kaolinite specimen, thus dampening it, causes that mineral to exude a musty or dank odor.

Feel

The feel of a mineral is the impression gained by handling or rubbing it. Terms used to describe feel are common descriptive adjectives such as *soapy, greasy, smooth, rough,* and so forth.

Chemical Reaction

Calcite will effervesce (bubble) when treated with cold dilute (1N) hydrochloric acid. NOTE: Your laboratory instructor will provide the proper dilute acid if you are to use this test.

Name *Section* *Date*

Exercise 1

Identification of Common Minerals

Your instructor will provide you with a variety of minerals to be identified. Take time to examine the minerals and review the various physical properties described in the previous pages. Select several samples and examine them for luster, color, hardness, and streak, and compare their specific gravity (G) using the heft test.

Study table 1.2. Note that certain minerals have physical properties that make their identification relatively easy. For example, graphite is soft, feels greasy, and marks both your hands and paper. Galena is "heavy," shiny, and has perfect cubic cleavage. Calcite has perfect rhombohedral cleavage, is easily scratched by a knife, reacts with cold dilute HCl, and in transparent specimens shows double refraction.

When you feel that you have an understanding of the various physical properties and the tests that you must apply to determine these properties, select a specimen at random from the group of minerals provided to you in the laboratory. Refer to figures 1.5 through 1.20 as an aid to identification. Be aware that your laboratory collection may contain some minerals that are not shown in the figures. Due to the normal variations within a single mineral species, some of the specimens may appear different from the same minerals shown in the figures in this manual.

Using the worksheets provided on the following pages to record your observations, follow the steps outlined below.

1. Examine carefully a single mineral specimen selected at random from the group provided to you in the laboratory.
2. Determine whether the sample has a metallic or nonmetallic luster. Then determine whether it is light- or dark-colored. (The terms *dark* and *light* are subjective. A mineral that is "dark" to one observer may be "light" to another. This possibility is anticipated in table 1.2 where mineral specimens that could fall into either the "light" or "dark" categories are listed in both groups. The same is true for minerals that may exhibit either metallic or nonmetallic luster.)
3. If the mineral falls into either group I or II, proceed to test it first for hardness and then for cleavage. This will place the mineral with a small group of other minerals in table 1.2. Identification can be completed by noting other diagnostic general or specific physical properties.

   If the mineral falls into group III, test it for streak and note other general and specific properties such as color, hardness, cleavage, and so on, until the mineral fits the description of one of those given in table 1.2 under group III.
4. To assist you in confirming your identification, refer to the expanded mineral descriptions in table 1.3.
5. Your laboratory instructor will advise you as to the procedure to be used to verify your identification.
6. Refer to table 1.3 to learn about occurrence, economic value, and uses of each mineral. The chemical groupings and composition of some of the common minerals are presented in table 1.4.

TABLE 1.2 *Mineral Identification Key*

Group	Hardness	Cleavage	Properties	Mineral
I. Nonmetallic, light-colored	Hard (scratches glass)	Shows cleavage	Vitreous luster. Color varies from white or cream to pink. Hardness 6. Cleavage two planes at nearly 90 degrees. Streak white. G = 2.6. Crystals common. Grains have glossy appearance.	ORTHOCLASE
			Vitreous luster. Color varies from white to gray or reddish to reddish brown. Hardness 6.0–6.5. Cleavage two planes at nearly 90 degrees. Cleavage planes show striations. Streak white. G = 2.6–2.8. Striations diagnostic. Some samples may show a play of colors.	PLAGIOCLASE
		No cleavage	Vitreous luster. Colorless or white, but almost any color can occur. Hardness 7. Cleavage none. Conchoidal fracture. Streak white. G = 2.65. Hexagonal crystals with striations perpendicular to the long dimension of the crystal common. Also massive.	QUARTZ
			Waxy or dull luster. Color varies from white to pale yellow, brown, or gray. Hardness 7. Cleavage none. Streak white. G = 2.6. Characterized by conchoidal fracture with sharp edges.	CHALCEDONY (flint/chert)
			Waxy luster. Variegated banded colors. Hardness 7. Cleavage none. Streak white. G = 2.6.	CHALCEDONY (agate)
			Vitreous luster. Color commonly olive-green, sometimes yellowish. Hardness 6.5–7. Cleavage indistinct. Streak white or gray. G = 3.2–4.4. Commonly in granular masses.	OLIVINE
	Soft (does not scratch glass)	Shows cleavage	Vitreous luster. Colorless, also white, gray, yellow, or red. Hardness 2.5. Perfect cubic cleavage. Streak white. G = 2.2. Table salt taste.	HALITE
			Vitreous luster. Colorless and transparent, white, variety of colors possible. Hardness 3. Perfect rhombohedral cleavage. Streak white to gray. G = 3.0. Effervesces in cold dilute HCl; double refraction in transparent varieties.	CALCITE
			Vitreous to pearly luster. Colorless, white, pink, gray, greenish, or yellow-brown. Hardness 3.5–4. Rhombohedral cleavage. Streak white. G = 2.9–3.0. Crystals common, twinning common. Reaction with cold dilute HCl only when powdered.	DOLOMITE
			Vitreous to pearly luster. Colorless to white, gray, yellowish orange, or light brown. Hardness 2. Cleavage good in one direction producing thin sheets. Fracture may be fibrous. Streak white. G = 2.3. Crystals common, twinning common.	GYPSUM (selenite)
			Pearly to greasy to dull luster. Color usually pale green, also shades of white or gray. Hardness 1. Perfect basal cleavage. Streak white. G = 2.6–2.8. Soapy feel.	TALC
			Vitreous to silky or pearly luster. Colorless to shades of green, gray, or brown. Hardness 2.5–3.5. Perfect basal cleavage yielding thin flexible and elastic sheets. Streak white. G = 2.8–2.9.	MUSCOVITE
			Greasy waxlike to silky luster. Color varies, shades of green most common. Hardness 2.0–3.0. Cleavage imperfect. Fibrous parting. Streak white. G = 2.5–2.6.	ASBESTOS (crysotile)
			Vitreous luster. Colorless but wide range of colors possible. Hardness 4. Perfect octahedral cleavage (4 planes). Streak white. G = 3.2.	FLUORITE
		No cleavage	Dull to earthy luster. Color white, often stained. Hardness 2.0–2.5. No cleavage apparent in common massive varieties. Streak white. G = 2.6–2.7. Earthy smell when damp.	KAOLINITE
			Pearly to greasy or dull luster. Color pale green or shades of gray. Hardness 1. No apparent cleavage in massive varieties. Streak white. G = 2.6–2.8. Soapy feel.	TALC
			Earthy luster. White or various colors. Hardness 3, may be less. No apparent cleavage in massive varieties. Streak white, G = 3.0. Effervesces in cold dilute HCl.	CALCITE
			Earthy luster. White or various colors. Hardness 3.5–4, but apparent may be less. No apparent cleavage. Streak white. G = 2.9–3.0. Reacts with cold dilute HCl only when powdered.	DOLOMITE
			Earthy luster. White color. Hardness 2, but apparent may be less. No apparent cleavage in massive varieties. Streak white. G = 2.3. Massive fine-grained variety called *alabaster*, fibrous variety called *satin spar.*	GYPSUM (alabaster) (satin spar)

TABLE 1.2 *Mineral Identification Key (Continued)*

Group	Hardness	Cleavage	Description	Mineral
II. Nonmetallic, dark-colored	Hard (scratches glass)	Shows cleavage	Vitreous luster. Color dark green to black. Hardness 5.5–6.0. Cleavage two planes at nearly 90 degrees. Streak white to gray. G = 3.2–3.6. May exhibit parting.	AUGITE
			Vitreous luster. Color dark green to black. Hardness 5–6. Cleavage two planes with intersections at 56 and 124 degrees. G = 3.0–3.5. Six-sided crystals common.	HORNBLENDE
			Vitreous luster. Color varies from white, gray, to reddish or reddish brown. Hardness 6.0–6.5. Cleavage two planes at nearly 90 degrees. Striations on cleavage planes. Streak white. G = 2.6–2.8. Some forms exhibit play of colors on cleavage surfaces.	PLAGIOCLASE
		No cleavage	Vitreous luster. Color varies but commonly brown. Hardness 9. Cleavage none. G = 4.0. Barrel-shaped hexagonal crystals with striations on basal faces common.	CORUNDUM
			Vitreous to resinous luster. Color varies but dark red to reddish brown common. Hardness 7.0–7.5. Cleavage none. Streak white or shade of the mineral color. G = 3.4–4.2. Fracture may resemble a poor cleavage. Brittle.	GARNET
			Vitreous luster. Color commonly olive-green, sometimes yellowish. Hardness 6.5–7.0. Cleavage indistinct. Streak white or gray. G = 3.2–4.4. Commonly in granular masses.	OLIVINE
			Vitreous luster. Color gray to gray-black. Hardness 7. Cleavage none. Streak white. G = 2.65. Conchoidal fracture. Crystals common; also a variety of massive forms.	QUARTZ
			Waxy to dull luster. Color red to red-brown or brown. Hardness 7. Cleavage none. Streak white to gray. G = 2.6.	CHALCEDONY (jasper)
			Waxy or dull luster. Color dark gray to black. Hardness 7. Cleavage none. Streak white. G = 2.6. Characterized by conchoidal fracture with sharp edges.	CHALCEDONY (flint/chert)
	Soft (does not scratch glass)	Shows cleavage	Vitreous to pearly luster. Color dark green, brown, to black. Hardness 2.5–3.0. Perfect basal cleavage forming thin elastic sheets. Streak white to gray. G = 2.7–3.4.	BIOTITE
			Resinous luster. Color yellow-brown to dark brown. Hardness 3.5–4.0. Cleavage perfect in six directions (dodecahedral). Streak brown to light yellow or white. G = 3.9–4.1. Cleavage faces common; twinning common.	SPHALERITE
			Vitreous to earthy luster. Color green to greenish black. Hardness 2.0–3.0. Perfect basal cleavage forming flexible nonelastic sheets. Streak white to pale green. G = 2.3–3.3. May have slippery feel.	CHLORITE
		No cleavage	Submetallic to earthy luster. Color red to red-brown. Hardness 5–6 but apparent may be as low as 1. Cleavage none. Streak red. G = 5.0–5.3. Earthy appearance.	HEMATITE (soft iron ore)
			Vitreous to subresinous luster. Color varies; green, blue, brown, purple. Hardness 5. Cleavage poor basal. Streak white. G = 3.1–3.2. Crystals common.	APATITE
			Earthy luster. Color varies; yellow, yellow-brown to brownish black. Apparent hardness 1. No cleavage apparent to earthy masses. Streak brownish yellow to orange-yellow. G = 3.3–4.3. Earthy masses.	GOETHITE (limonite)

TABLE 1.2 *Mineral Identification Key (Continued)*

Luster	Streak	Description	Mineral
III. Metallic luster	Black, green-black or dark green streak	Metallic luster. Color black. Hardness 5.5–6.0. Cleavage none. Streak black. G = 5.2. Strongly magnetic.	MAGNETITE
		Metallic luster. Color dark gray to black. Hardness 1.0. Perfect basal cleavage. Streak black. G = 2.1–2.2. Greasy feel, smudges fingers when handled.	GRAPHITE
		Metallic luster. Color brass-yellow. Hardness 6–6.5. Cleavage none. Streak greenish or brownish black. G = 4.8–5.0. Cubic crystals with striated faces common. "*Fools gold.*"	PYRITE
		Metallic luster. Color brass-yellow, often tarnished to bronze or purple. Hardness 3.5–4. Cleavage none. Streak greenish black. G = 4.1–4.3. Usually massive.	CHALCOPYRITE
		Bright metallic luster. Color shiny lead-gray. Hardness 2.5. Perfect cubic cleavage. Streak lead-gray. G = 7.6. Cleavage, high G, and softness diagnostic.	GALENA
	Red streak	Metallic luster. Color steel-gray. Hardness 5–6. Cleavage none. Streak red to red-brown. G = 5.3. Often micaceous or foliated. Brittle.	HEMATITE (specularite)
	Yellow, brown, or white streak	Metallic to dull luster. Color yellow-brown to dark brown, may be almost black. Hardness 5.0–5.5. Cleavage perfect parallel to side pinacoid. Streak brownish yellow to orange-yellow. G = 4.3. Brittle.	GOETHITE
		Submetallic to resinous luster. Color yellow to yellow-brown to dark brown. Hardness 3.5–4. Perfect cleavage in six directions (dodecahedral). Streak brown to light yellow to white. G = 3.9–4.1. Cleavage faces common; twinning common.	SPHALERITE

Note: Values for hardness and specific gravity have in most cases been rounded to the nearest tenth and have been revised to reflect the data in Dana's New Mineralogy, *8th Edition, Richard V. Gaines, H. Catherine W. Skinner, Eugene E. Foord, Bryan Mason, and Abraham Rosenzweig, John Wiley and Sons, Inc., New York, 1997.*

FIGURE 1.5 Quartz crystal.

FIGURE 1.6 Rose quartz.

FIGURE 1.7 Smoky quartz.

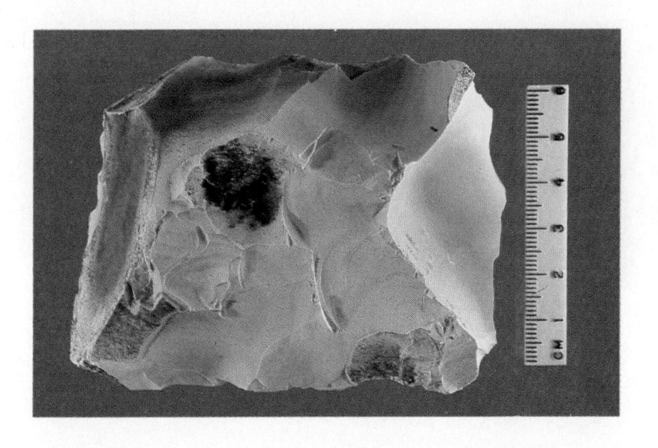

FIGURE 1.8 Crypotocrystalline quartz (chert).

FIGURE 1.9 Orthoclase (microcline).

FIGURE 1.10 Plagioclase.

FIGURE 1.11 Gypsum (selenite).

FIGURE 1.12 Talc.

FIGURE 1.13 Calcite (note double refraction).

FIGURE 1.14 Fluorite.

FIGURE 1.15 Biotite.

FIGURE 1.16 Olivine.

FIGURE 1.17 Hematite (specularite).

FIGURE 1.18 Hematite.

FIGURE 1.19 Goethite (limonite).

FIGURE 1.20 Pyrite.

TABLE 1.3 *Mineral Catalog*

Mineral	Description
APATITE Ca, F Phosphate Hexagonal	**Luster:** nonmetallic; vitreous to subresinous. **Color:** varies, greenish yellow, blue, green, brown, purple, white. **Hardness:** 5. **Cleavage:** poor basal; fracture conchoidal. **Streak:** white. **G** = 3.1–3.2. **Habit:** crystals common. Also, massive or granular forms. **Uses:** important source of phosphorus for fertilizers, phosphoric acid, detergents, and munitions.
ASBESTOS (Serpentine) Mg, Al Silicate Monoclinic	**Luster:** nonmetallic; greasy or waxlike in massive varieties, silky when fibrous. **Color:** varies, shades of green most common. **Hardness:** 2.0–3.0. **Cleavage:** none. **Streak:** white. **G** = 2.5–2.6. **Habit:** occurs in massive, platy, and fibrous forms. **Uses:** wide industrial uses, especially in roofing materials.
AUGITE Ca, Mg, Fe, Al Silicate Monoclinic	**Luster:** nonmetallic; vitreous. **Color:** dark green to black. **Hardness:** 5.5–6.0. **Cleavage:** good, two planes at nearly 90°; may exhibit well-developed parting. **Streak:** white to gray. **G** = 3.2–3.6. **Habit:** short, stubby eight-sided prismatic crystals. Often in granular crystalline masses. Most important ferromagnesium mineral in dark igneous rocks. **Uses:** no commercial value. *NOTE:* The *pyroxene* group of minerals contains over 15 members. Augite is an example of the monoclinic members of this group.
BIOTITE K, Mg, Fe, Al Silicate Monoclinic	**Luster:** nonmetallic; vitreous to pearly. **Color:** dark green, brown, or black. **Hardness:** 2.5–3.0. **Cleavage:** perfect basal forming elastic sheets. **Streak:** white to gray. **G** = 2.7–3.4. **Habit:** crystals common as pseudohexagonal prisms, more commonly in sheets or granular crystalline masses. Common accessory mineral in igneous rocks, also important in some metamorphic rocks. **Uses:** commercial value as insulator and in electrical devices.
CALCITE $CaCO_3$ Trigonal	**Luster:** nonmetallic; vitreous (may appear earthy in fine-grained massive forms). **Color:** colorless and transparent or white when pure; wide range of colors possible. **Hardness:** 3. **Cleavage:** perfect rhombohedral. **Streak:** white to gray. **G** = 2.7. **Habit:** crystals common. Also massive, granular, oolitic, or in a variety of other habits. Effervesces in cold dilute HCl. Strong double refraction in transparent varieties. Common and widely distributed rock-forming mineral in sedimentary and metamorphic rocks. **Uses:** chief raw material for portland cement; wide variety of other uses.
CHALCEDONY (Cryptocrystalline Quartz) SiO_2 Trigonal	**Luster:** nonmetallic; waxy or dull. **Color:** varies. Chalcedony is a group name for a variety of extremely fine-grained, very diverse forms of quartz including *agate,* banded forms; *carnelian* or *sard,* red to brown; *jasper,* opaque and generally red or brown; *chert* and *flint,* massive, opaque and ranging in color from white, pale yellow, brown, gray, to black and characterized by conchoidal fracture with sharp edges; *silicified wood,* reddish or brown showing wood structure. **Hardness:** 7. **Cleavage:** none. **Streak:** white. **G** = 2.6. **Habit:** occurs in a wide variety of sedimentary rocks and veins, cavities, or as dripstone. **Uses:** various forms used as semiprecious stones.
CHALCOPYRITE $CuFeS_2$ Trigonal	**Luster:** metallic, opaque. **Color:** brass-yellow, often tarnished to bronze or purple. **Hardness:** 3.5–4. **Cleavage:** none, fracture uneven. **Streak:** greenish black. **G** = 4.1–4.3. **Habit:** may occur as small crystals but usually massive. **Uses:** most important as common copper ore mineral.
CHERT (see Chalcedony)	
CHLORITE Mg, Fe, Al Silicate Monoclinic	**Luster:** nonmetallic; vitreous to earthy. **Color:** green to green-blue. **Hardness:** 2.5–3.0. **Cleavage:** perfect, basal forming flexible nonelastic sheets. **Streak:** white to pale green. **G** = 2.3–3.3. **Habit:** occurs as foliated masses or small flakes. Common in low-grade schists and as alteration product of other ferromagnesian minerals. **Uses:** no commercial value.

TABLE 1.3 *Mineral Catalog (Continued)*

Mineral	Properties
CORUNDUM Al_2O_3 Trigonal	**Luster:** nonmetallic, adamantine to vitreous. **Color:** varies; yellow, brown, green, purple; gem varieties blue (sapphire) and red (ruby). **Hardness:** 9. **Cleavage:** none, basal parting common with striations on parting planes. **Streak:** white, **G** = 4. **Habit:** barrel-shaped crystals common, frequently with deep horizontal striations. **Uses:** wide uses as an abrasive.
DOLOMITE $CaMg(CO_3)_2$ Trigonal	**Luster:** nonmetallic, vitreous to pearly. **Color:** colorless, white, gray, greenish, yellow-brown; other colors possible. **Hardness:** 3.5–4. **Cleavage:** rhombohedral. **Streak:** white. **G** = 2.9–3.0. **Habit:** crystals common. Twinning common. Fine-grained; massive and granular forms common also. Distinguished from calcite by fact that it effervesces in cold dilute HCl only when powdered. Widespread occurrence in sedimentary rocks. **Uses:** variety of industrial uses as flux, as source of magnesia for refractory bricks, and as a source of magnesium or calcium metal.
FLINT/JASPER (see Chalcedony)	
FLUORITE CaF_2 Cubic (Isometric)	**Luster:** nonmetallic, vitreous transparent to translucent. **Color:** colorless when pure; occurs in a wide variety of colors: yellow, green, blue, purple, brown, and shades in between. **Hardness:** 4. **Cleavage:** perfect octahedral (four directions parallel to faces of an octahedron). **Streak:** white. **G** = 3.2. **Habit:** twins fairly common. Common as a vein mineral. **Uses:** industrial use as a flux in steel and aluminum metal smelting; source of fluorine for hydrofluoric acid.
GALENA PbS Cubic (Isometric)	**Luster:** bright metallic. **Color:** lead-gray. **Hardness:** 2.5. **Cleavage:** perfect cubic. **Streak:** lead-gray. **G** = 7.6. **Habit:** crystals common, easily identified by cleavage, high specific gravity, and softness. **Uses:** chief lead ore.
GARNET Fe, Mg, Ca, Al Silicate Cubic (Isometric)	**Luster:** nonmetallic; vitreous to resinous. **Color:** varies but dark red and reddish brown most common; white, pink, yellow, green, black depending on composition. **Hardness:** 7.0–7.5. **Cleavage:** none. **Streak:** white or shade of mineral color. **G** = 3.4–4.2. **Habit:** crystals common. Also in granular masses. **Uses:** some value as an abrasive. Gemstone varieties are pyrope (red) and andradite (green).
GOETHITE FeO (OH) Orthorhombic	**Luster:** varies, crystals adamantine; metallic to dull in masses, fibrous varieties may be silky. **Color:** dark brown, yellow-brown, reddish brown, brownish black, yellow. **Hardness:** 5–5.5. **Cleavage:** perfect parallel to side pinacoid, fracture uneven. **Streak:** brownish yellow to orange-yellow. **G** = 4.3. **Habit:** crystals uncommon. Usually massive or earthy as residual from chemical weathering, or stalactic by direct precipitation. Often in radiating fibrous forms. This species includes the common brown and yellow-brown ferric oxides collectively called *limonite.* **Uses:** commercial source of iron ore.
GRAPHITE C Hexagonal	**Luster:** metallic to dull. **Color:** dark gray to black. **Hardness:** 1.0. **Cleavage:** perfect basal. **Streak:** black. **G** = 2.1–2.2. **Habit:** characteristic greasy feel, marks easily on paper. Crystals uncommon, usually as foliated masses. Common metamorphic mineral. **Uses:** wide industrial uses due to high melting temperature (3000°C) and insolubility in acid. Used also as lead in pencils.
GYPSUM $CaSO_4{\cdot}2H_2O$ Monoclinic	**Luster:** nonmetallic; vitreous to pearly; some varieties silky. **Color:** colorless to white, gray, yellowish orange, light brown. **Hardness:** 2. **Cleavage:** good in one direction producing thin sheets; fracture conchoidal in one direction, fibrous in another. **Streak:** white. **G** = 2.3. **Habit:** crystals common and simple in habit; twinning common. Varieties include *selenite,* coarsely crystalline, colorless and transparent; *satinspar,* parallel fibrous structure; and *alabaster,* massive fine-grained gypsum. Occurs widely as sedimentary deposits and in many other ways. **Uses:** mined for use in wallboard, plaster, and filler for paper products.

TABLE 1.3 *Mineral Catalog (Continued)*

HALITE $NaCl$ Cubic (Isometric)	**Luster:** nonmetallic, vitreous. Transparent to translucent. **Color:** colorless, also white, gray, yellow, red. **Hardness:** 2.5. **Cleavage:** perfect cubic. **Streak:** white. **G** = 2.2. **Habit:** crystals common, also massive or coarsely granular. Characteristic taste of table salt. **Uses:** widely used as source of both sodium and chlorine and as salt for table, pottery glaze, and industrial purposes.
HEMATITE Fe_2O_3 Trigonal	**Luster:** metallic in form known as specularite and in crystals; submetallic to dull in other varieties. **Color:** steel-gray in specularite, dull to bright red in other varieties. **Hardness:** 5–6, but apparent may be as low as 1. **Cleavage:** none; basal parting fracture uneven. **Streak:** red-brown. **G** = 5.3. **Habit:** crystals uncommon. May occur in crystalline, botryoidal, or earthy masses. Specularite commonly micaceous or foliated. **Streak:** characteristic. **Uses:** common iron ore.
HORNBLENDE (Amphibole) Ca, Na, Mg, Fe, Al Silicate Monoclinic	**Luster:** nonmetallic, vitreous. **Color:** dark green, dark brown, black. **Hardness:** 5.0–6.0 **Cleavage:** perfect on two planes meeting at 56° and 124°. **Streak:** gray or pale green. **G** = 3.0–3.5. **Habit:** long, six-sided crystals common. **Color:** usually darker than other minerals in amphibole group. *Tremolite-Actinolite,* also a member of the amphibole group, is lighter in color and commonly is fibrous or asbestiform, may range in color from white to green, and has white streak. **Uses:** no commercial use. *Nephrite,* the amphibole form of jade, is used for jewelry.
KAOLINITE Hydrous Aluminosilicate Triclinic	**Luster:** nonmetallic, dull to earthy. **Color:** white, often stained by impurities to red, brown, or gray. **Hardness:** 2.0–2.5. **Cleavage:** perfect basal but rarely seen because of small grain size. **Streak:** white. **G** = 2.6–2.7. **Habit:** found in earthy masses. Earthy odor when damp. *NOTE:* Kaolinite is used here as an example of the clay minerals. It normally is not possible to distinguish the various clay minerals on the basis of their physical properties. Other clay minerals include *montmorillonite* (smectite), *illite,* and *vermiculite.* **Uses:** kaolinite is used as a paper filler and ceramic, montmorillonite for drilling muds, illite has no industrial use, and vermiculite is mined and processed for use as a lightweight aggregate, potting soils, and as insulation.
LIMONITE (see Goethite)	
MAGNETITE $FeFe_2O_4$ some Cubic (Isometric)	**Luster:** metallic. **Color:** black. **Hardness:** 5.5–6.0. **Cleavage:** none, some octahedral parting. **Streak:** black. **G** = 5.2. **Habit:** usually in granular masses. Strongly magnetic, specimens show polarity (lodestones). Widespread occurrence in a variety of rocks. **Uses:** used commercially as iron ore.
MUSCOVITE K, Al Silicate Monoclinic	**Luster:** nonmetallic, vitreous to silky or pearly. **Color:** colorless to shades of green, gray, or brown. **Hardness:** 2.5–3.5. **Cleavage:** perfect basal yielding thin sheets that are flexible and elastic; may show some parting. **Streak:** white. **G** = 2.8–2.9. **Habit:** usually in small flakes or lamellar masses. Commercial deposits found in granite pegmatites but occur in many other rocks. **Uses:** variety of industrial uses.
OLIVINE $(Mg, Fe)_2SiO_4$ Orthorhombic	**Luster:** nonmetallic, vitreous. **Color:** olive-green to yellowish; nearly pure Mg-rich varieties may be white (forsterite) and nearly pure Fe-rich varieties brown to black (fayalite). **Hardness:** 6.5–7. **Cleavage:** indistinct. **Streak:** white or gray. **G** = 3.2–4.4. **Habit:** usually in granular masses. Crystals uncommon. A mineral of basic and ultrabasic rocks. **Uses:** forsterite variety used for refractory bricks.

TABLE 1.3 *Mineral Catalog (Continued)*

ORTHOCLASE (K-Feldspar) $K(AlSi_3O_8)$ Monoclinic	**Luster:** nonmetallic, vitreous. **Color:** varies, white, cream, or pink; *sanidine* variety may be colorless. **Hardness:** 6. **Cleavage:** two planes at nearly right angles. **Streak:** white. **G** = 2.6. **Habit:** crystals not common. Has glossy appearance. Distinguished from other feldspars by absence of twinning striations. *NOTE: Microcline* variety is triclinic. When light green the color is diagnostic, more commonly white, green, pink. Occurrence helpful; most K-feldspar in pegmatites is microcline. **Uses:** commonly used in ceramics, glassmaking, and in scouring and cleansing products.
PLAGIOCLASE Ranges in composition from Albite, $NaAlSi_3O_8$, to Anorthite, $CaAl_2Si_2O_8$ Triclinic	**Luster:** nonmetallic, vitreous. **Color:** white or gray, reddish, or reddish brown. **Hardness:** 6.0–6.5. **Cleavage:** two planes at close to right angles, twinning striations common on basal cleavage surfaces. **Streak:** white. **G** = 2.6–2.8. **Habit:** crystals common for Na-rich varieties, uncommon for intermediate varieties, rare for anorthite. Twinning common. Twinning striations on basal cleavage useful to distinguish from orthoclase. Some varieties show play of colors. **Uses:** sodium-rich varieties mined for use in ceramics.
PYRITE FeS_2 Cubic (Isometric)	**Luster:** metallic. **Color:** brass-yellow, may be iridescent if tarnished. **Hardness:** 6–6.5. **Cleavage:** none, conchoidal fracture. **Streak:** greenish or brownish black. **G** = 4.8–5.0. **Habit:** crystals common. Usually cubic with striated faces. Crystals may be deformed. Massive granular forms also. Most widespread sulfide mineral. Known as "fool's gold." **Uses:** source of sulfur for sulfuric acid. *NOTE: Marcasite* (FeS_2) is orthorhombic, usually paler in color, and is commonly altered.
QUARTZ SiO_2 Trigonal	**Luster:** nonmetallic, vitreous. **Color:** typically colorless or white, but almost any color may occur. **Hardness:** 7. **Cleavage:** none, conchoidal fracture. **Streak:** white but difficult to obtain on streak plate. **G** = 2.65. **Habit:** prismatic crystals common with striations perpendicular to the long dimension; also a variety of massive forms. Color variations lead to varieties called smoky quartz, rose quartz, milky quartz, and amethyst. Common mineral in all categories of rocks. **Uses:** wide variety of commercial uses including glassmaking, electronics, and in construction products.
SPHALERITE ZnS Cubic (Isometric)	**Luster:** usually nonmetallic, some varieties submetallic, most commonly resinous. **Color:** yellow, yellow-brown to dark brown. **Hardness:** 3.5–4. **Cleavage:** perfect dodecahedral (six directions at 120°). **Streak:** brown to light yellow or white. **G** = 3.9–4.1. **Habit:** crystals common as distorted tetrahedra or dodecahedra. Twinning common. Also massive or granular. **Uses:** important zinc ore.
TALC Mg Silicate Monoclinic	**Luster:** nonmetallic, pearly to greasy or dull. **Color:** usually pale green, also white to silver-white or gray. **Hardness:** 1. **Cleavage:** perfect basal, massive forms show no visible cleavage. **Streak:** white. **G** = 2.6–2.8. Usually foliated masses or dense fine-grained dark gray to green aggregates (soapstone). **Habit:** crystals extremely rare. Soapy feel is diagnostic. **Uses:** commercial uses in paints, ceramics, roofing, paper, and toilet articles.

TABLE 1.4 *Chemical Grouping and Composition of Some Common Minerals*

	EXAMPLE	
CHEMICAL GROUP	MINERAL NAME	CHEMICAL FORMULA*†
ELEMENTS	Native copper	Cu
	Graphite	C
	Diamond	C
OXIDES	Quartz	SiO_2
	Hematite	Fe_2O_3
	Magnetite	Fe_3O_4
	Goethite	FeO(OH)
	Corundum	Al_2O_3
SULFIDES	Pyrite	FeS_2
	Chalcopyrite	$CuFeS_2$
	Galena	PbS
	Sphalerite	ZnS
SULFATES	Anhydrite	$CaSO_4$
	Gypsum	$CaSO_4 \cdot 2H_2O$
CARBONATES	Calcite	$CaCO_3$
	Dolomite	$CaMg(CO_3)_2$
PHOSPHATES	Apatite	$Ca_5(PO_4)_3F$
HALIDES	Halite	NaCl
	Fluorite	CaF_2
SILICATES: OLIVINE GROUP	Olivine	$(Mg,Fe)_2SiO_4$
SILICATES: AMPHIBOLE GROUP	Hornblende	Ca,Na,Mg,Fe,Al Silicate
	Asbestos (fibrous serpentine)	Mg,Al Silicate
SILICATES: PYROXENE GROUP	Augite	Ca,Mg,Fe,Al Silicate
SILICATES: MICA GROUP	Muscovite Biotite Chlorite Talc Kaolinite	K,Al Silicate K,Mg,Fe,Al Silicate Mg,Fe,Al Silicate Mg Silicate Al Silicate
SILICATES: FELDSPAR GROUP	Orthoclase (K-feldspar)	$K(AlSi_3O_8)$
	Plagioclase (Ab,An)	Mixture of Ab and An
	Albite (Ab)	$NaAlSi_3O_8$
	Anorthite (An)	$CaAl_2Si_2O_8$
SILICATES: GARNET GROUP	Garnet	Fe,Mg,Ca,Al Silicate

*Some common elements and their symbols:
Al-Aluminum, C-Carbon, Ca-Calcium, Cl-Chlorine, Cu-Copper, F-Fluorine, Fe-Iron, H-Hydrogen, K-Potassium, Mg-Magnesium, Na-Sodium, O-Oxygen, P-Phosphorus, Pb-Lead, S-Sulfur, Si-Silicon, Zn-Zinc

†Chemical formulas from *Mineralogy,* 2d ed., by L. G. Berry, Brian Mason, and R. V. Dietrich (San Francisco: W. H. Freeman, 1983).

WEB CONNECTIONS

MINERALS

http://minerals.er.usgs.gov/minerals/

Statistics and information on the worldwide supply, demand, and flow of minerals and materials essential to the U.S. economy, the national security, and protection of the environment.

http://web.wt.net/~data/Mineral/

A mineralogy database in HTML format. Information on crystallography, chemical composition, physical and optical properties, classification, and alphabetical listings of over 3,800 minerals. Links to other sources of mineralogy data are available at this site.

http://un2sg4.unige.ch/athena/mineral/mineral.html

Mineral lists of IMA approved mineral names and varieties names. Classification, crystal system, etc. and some images are also included.

http://mineral.galleries.com/

A collection of mineral descriptions, images, and specimens, continuously updated. Minerals are listed by name, class, interesting groupings such as gemstones, and there is a system of keyword searching for mineral identification.

http://www.uni-wuerzburg.de/mineralogie/links.html

A source of links on the Internet to a wide variety of mineralogical and geologic sites that may be of interest to mineralogists. The subjects are quite broad and should be of interest to all geologists.

http://www.iumsc.indiana.edu/doc/crystmin.html

An explanation of the cubic crystal system and some of the classes within it. Allows students to rotate applets to see symmetry elements and location in various forms.

http://galaxy.einet.net/images/gems/gems-icons.html

The Smithsonian gem and mineral collection is featured at this site with both images and descriptions. For those who are interested, a "don't miss" site on the web.

Name Section Date

Worksheet for Minerals

Sample #	Luster	Color Lgt/Dk	Hardness	Cleavage Angles	Streak	Special Properties?	Mineral Name	Chemical Composition

Name Section Date

Worksheet for Minerals

Sample #	Luster	Color Lgt/Dk	Hardness	Cleavage Angles	Streak	Special Properties?	Mineral Name	Chemical Composition

ROCKS

BACKGROUND

As a part of the study of rocks in the lab or in the field, some understanding of the origin and occurrence of the rocks being studied is essential. The first part of this section deals with the identification and classification of rocks based on the study of hand specimens with reference also to their origin and occurrence. The second part of the section introduces some basic principles dealing with the relationships between rock types in their natural setting. These principles will be used in understanding the relative ages of adjacent rock masses; thus the relationships between rocks can help in tracing the geologic history of a given area. Several simple diagrams and schematic drawings will be used to convey the ideas and concepts.

Three major categories of rocks are recognized. They are *igneous,* rocks formed by the cooling and crystallization of molten material within or at the surface of the earth; *sedimentary,* rocks formed from sediments derived from preexisting rocks, by precipitation from solution, or by the accumulation of organic materials; and *metamorphic,* rocks resulting from the change of preexisting rocks into new rocks with different textures and mineralogy as a result of the effects of heat, pressure, chemical action, or combinations of these. All types of these three categories of rocks can be observed at the earth's surface today. We can also observe some of the processes that lead to their formation. For example, the eruption of volcanoes produces certain types of igneous rocks. We can observe the weathering, erosion, transportation, and deposition of sediments that upon *lithification*—the combination of processes (including compaction and cementation) that convert a sediment into an indurated rock—produce certain types of sedimentary rocks. Other observable processes of rock formation at the surface of the earth include the deposition of dripstone in caves and the growth of coral reefs in the oceans.

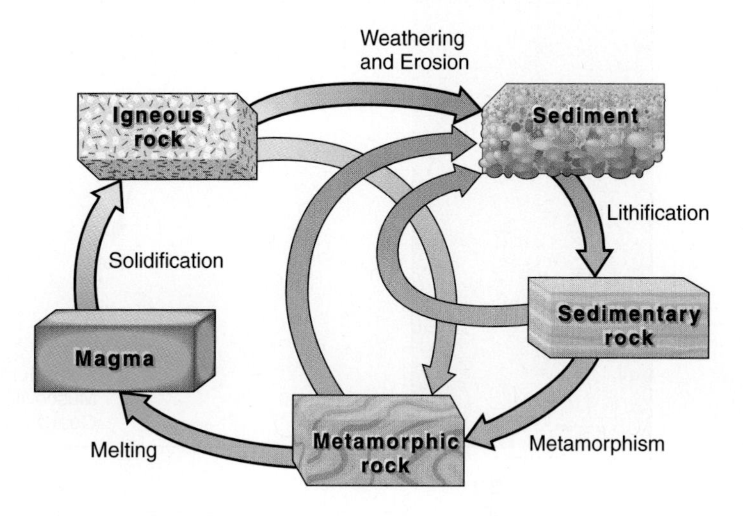

FIGURE 1.21

The rock cycle (shown schematically).

From Charles C. Plummer and David McGeary, Physical Geology, *8th edition.*

The earth's crust is dynamic and is subject to a variety of processes that act upon all types of rocks. Given time and the effect of these processes, any one of these rocks can be changed into another type. This relationship is the basis of the *rock cycle,* represented in figure 1.21. The heavy arrows indicate the normal complete cycle, while the open arrows indicate how this cycle may be interrupted. Keep in mind that the schematic presented in figure 1.21 is greatly simplified and that a given rock observed today represents only the last phase of the cycle from which the rock has been derived.

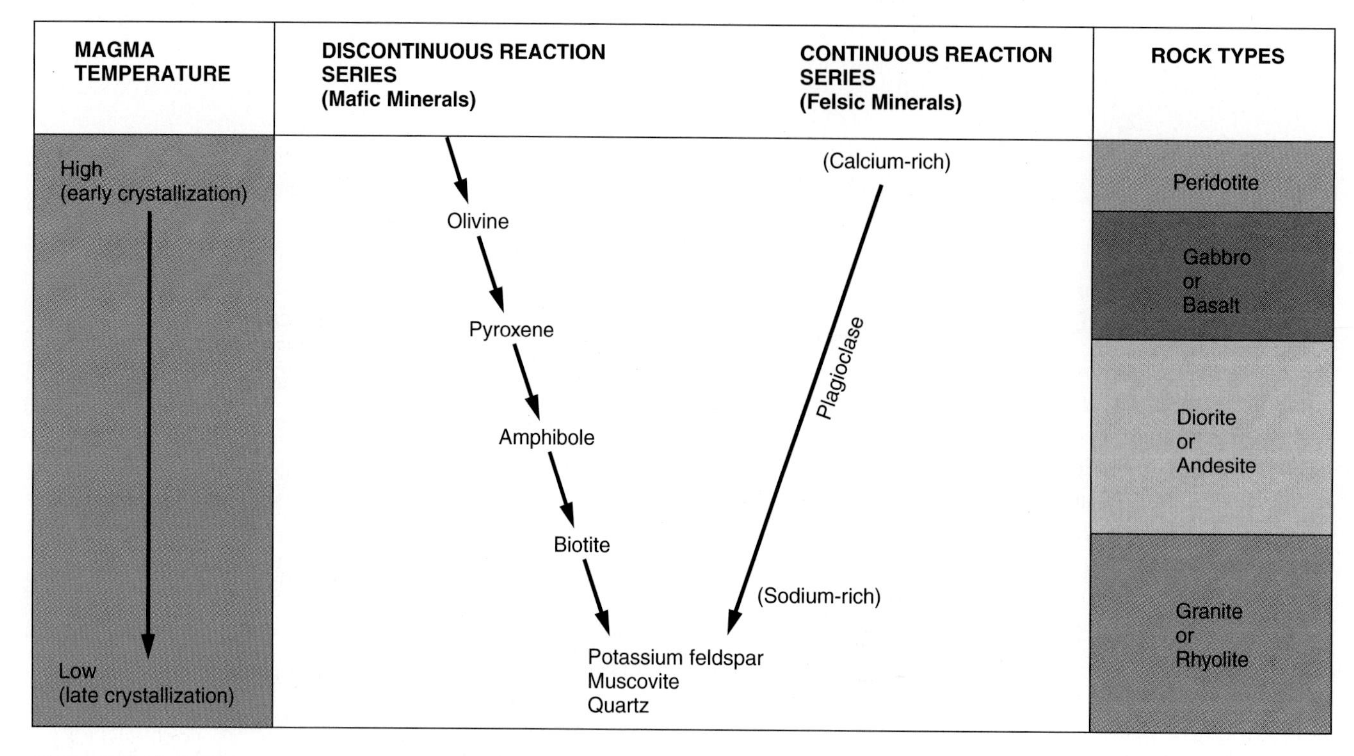

FIGURE 1.22

Reaction series for igneous rock formation from a magma.

Igneous Rocks

Origin

Igneous rocks are aggregates of minerals that crystallize from a molten material that is generated deep within the earth's mantle. The heat required to generate this melted material comes from within the earth. The temperature of the earth increases with depth at a rate of about 30° C per kilometer of depth. This increase in temperature is known as the *geothermal gradient,* and while the geothermal gradient varies from place to place, at a depth of about 35 kilometers the temperature is sufficient to melt rock. The molten material, *magma,* is a complex solution of silicates plus water and various gases. Some of this magma may reach the surface of the earth where it is extruded as *lava,* but other magmas solidify before they reach the surface. The rocks formed by solidification of magma within the mantle or crust are called *intrusive* igneous rocks, and those that form at the surface from lavas are called *extrusive* igneous rocks.

The composition of a given magma depends on the composition of the rocks that were melted to form the magma. Once melt occurs, the magma tends to rise toward the surface of the earth. As it rises, cooling and crystallization begin. Ultimately, the molten mass solidifies into solid rock. The type of igneous rock formed depends on a number of factors including the original composition of the magma, the rate of cooling, and the reactions that occurred within the magma as cooling took place.

Intrusive rock masses have been studied in detail, and it is recognized that there is an orderly sequence in which the various minerals crystallize. This sequence of crystallization (commonly called *Bowen's reaction series* after the geologist who first proposed it) has been verified in principle.

The reaction series presented in figure 1.22 gives some suggestion as to the formation of the various types of igneous rocks, and it helps to explain why the association of some minerals (such as olivine and quartz) is rare in nature. Study of figure 1.22 indicates that the first minerals to form tend to be low in silica. The plagioclase side of the diagram is labeled as a *continuous reaction series.* This means that the first crystals to form, calcium-rich plagioclase (anorthite), continue to react with the remaining liquid as cooling continues. This process involves the substitution of sodium for calcium in the plagioclase without a change in the crystal structure. As this process continues, the composition of the plagioclase becomes increasingly sodium- and silica-rich, and the last plagioclase formed is albite.

The *discontinuous reaction series* consists of the common ferromagnesian minerals found in igneous rocks. In this series the first mineral to crystallize is olivine. As the

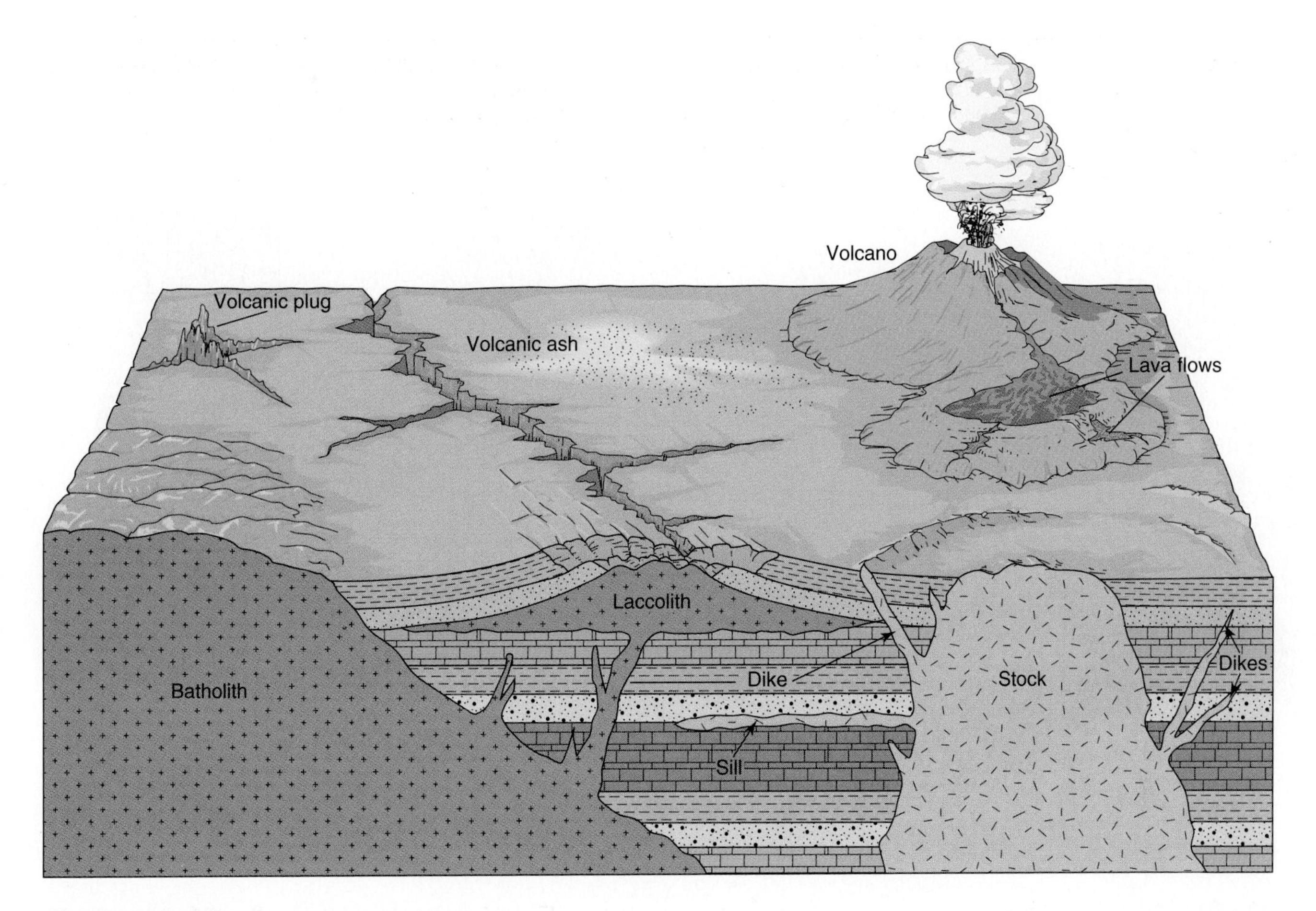

FIGURE 1.23

Block diagram showing various modes of occurrence of igneous rocks.

cooling continues, the olivine crystals react with the remaining liquid and form pyroxenes at the expense of olivine. As this process of interaction between the crystals and the silica-rich liquid continues, the pyroxene reacts to form amphibole, and the final member of this discontinuous series is biotite, which forms at the expense of amphibole.

If in fact the original magma is low in silica and high in iron and magnesium, the magma may solidify before the complete series of reactions has occurred. The resulting rocks are high in magnesium and iron, low in silica, and are said to be *mafic.* Olivine, pyroxene, and calcium-rich plagioclase are the common mineral associations of mafic rocks. Conversely, magma originally high in silica and low in ferromagnesian elements may reach the final stages of the reaction series, and the rocks formed are composed of potassium feldspar, quartz, and muscovite. These rocks are said to be *felsic* or *sialic.*

Keep in mind that this reaction series is idealized. Changes in the original composition of the magma occur in nature by *fractional crystallization*—the removal of crystals from the liquid magma by settling or filtering—by *assimilation* of part of the country rock through which the magma is rising, and/or by the *mixing* of two magmas of differing composition.

Occurrence

As magma works its way toward the surface of the earth, it encounters preexisting rock. The rock mass intruded by the magma is referred to as *country rock,* and it is not uncommon for pieces of the country rock to be engulfed by the magma. Fragments of country rock surrounded by igneous rocks are called *inclusions.*

Intrusive and extrusive igneous rocks may assume any number of geometric forms. Figure 1.23 illustrates several of the more common shapes of igneous rock masses found in nature. Table 1.5 shows the relationship of igneous rock types to their mode of occurrence.

The largest intrusive igneous rock mass (*pluton*) is a *batholith.* Batholiths are usually coarse-grained (*phaneritic*) in texture and granitic in composition. By definition, a batholith crops out over an area of more than 100 square kilometers (about 36 square miles). A *stock* is

FIGURE 1.24

Basaltic dike intruding Precambrian granite north of Lake Superior on the Canadian Shield.

TABLE 1.5 *Relationship of Igneous Rock Types to Their Modes of Occurrence in the Earth's Crust*

	ROCK TYPES	SOME MODES OF OCCURRENCE
EXTRUSIVE	Pumice Scoria	Pyroclastics Crusts on lava flows, pyroclastics
EXTRUSIVE	Obsidian	Lava flows
EXTRUSIVE	Rhyolite Andesite Basalt	Lava flows. (Also shallow intrusives such as dikes and sills)
INTRUSIVE	Rhyolite porphyry Andesite porphyry Basalt porphyry	Dikes, sills, laccoliths, intruded at medium to shallow depths
INTRUSIVE	Granite Diorite Gabbro Peridotite	Batholiths and stocks of deep-seated intrusive origin

similar in composition but, by definition, crops out over an area less than 100 kilometers (less than 36 square miles). A *dike* is a tabular igneous intrusion whose contacts cut across the trend of the country rock (fig. 1.24). A *sill* is also tabular in shape, but its contacts lie parallel to the trend of the country rock. A *laccolith* is similar to a sill but is generally much thicker, especially near its center where it has caused the country rock to bulge upward. These five major types of plutons—batholiths, stocks, dikes, sills, and laccoliths—can be grouped into two main categories based on the relationship of their contacts to the trend of the enclosing country rock. The contacts of batholiths, stocks, and dikes all cut across the trend of the country rock and hence are called *discordant igneous plutons.* Sills and laccoliths, on the other hand, have contacts that are parallel to the trends of the country rock and are called *concordant igneous plutons.* Table 1.6 summarizes these relationships.

TABLE 1.6 *Relationship between Concordant or Discordant Igneous Bodies and Their Modes of Occurrence*

RELATIONSHIP OF IGNEOUS ROCK CONTACT TO COUNTRY ROCK	MODE OF OCCURRENCE
Concordant	Laccolith, sill
Discordant	Batholith, stock, dike

TEXTURES OF IGNEOUS ROCKS

The *texture* of a rock is its appearance that results from the size, shape, and arrangement of the mineral grains or crystals in the rock. The texture of igneous rocks can be described in terms of one of the following:

Phaneritic or Coarse-Grained

This term applies to an igneous rock in which the constituent minerals are macroscopic in size. The dimensions of the individual crystals or grains range from about 1 mm to more than 5 mm (figs. 1.25–1.29).

Aphanitic or Fine-Grained

This term is used to describe the texture of an igneous rock composed of mineral crystals or grains that are microscopic in size; that is, they cannot be discerned with the naked or corrected eye (fig. 1.30).

Porphyritic

This term applies to an igneous rock in which the macroscopic mineral crystals or grains are embedded in a matrix of microscopic crystals or grains (figs. 1.32 and 1.34), or macroscopic crystals or grains of one size range occur in a matrix of smaller macroscopic crystals or grains. The larger crystals or grains are called *phenocrysts,* and the collection of smaller grains in which they are embedded is called the *groundmass.*

Vesicular

This texture is characterized by the presence of *vesicles*—tubular, ovoid, or spherical cavities in the rock (fig. 1.31). These "holes" in the rock are a result of gas bubbles being trapped in the rock as it cools. The size of the vesicles ranges from less than 1 mm up to several centimeters in diameter. Vesicles that are filled with mineral matter are called *amygdules,* and the texture then is called *amygdaloidal.*

Glassy

This texture resembles that of glass (fig. 1.39).

The texture of igneous rocks is a reflection of the mineralogy and the cooling history of the magma or lava from which they were formed. Igneous rocks with a phaneritic or coarse-grained texture were formed where the cooling rate was very slow and larger crystals or grains could form. Igneous rocks with an aphanitic or fine-grained texture were formed where cooling was much more rapid. A porphyritic texture reflects a two-stage cooling history. The larger crystals formed first under conditions of slow cooling, but before the magma or lava turned to rock it migrated to a zone of faster cooling where the remainder of the melt solidified. Vesicular textures are indicative of lava flows in which trapped gases produced the vesicles while the lava was still molten. A glassy texture reflects an extremely rapid rate of cooling. The surfaces of lava flows commonly possess a glassy texture.

MINERALOGIC COMPOSITION OF IGNEOUS ROCKS

The minerals of igneous rocks are grouped into two major categories, *primary* and *secondary.* Primary minerals are those that are crystallized from the cooling magma. Secondary minerals are those formed after the magma has solidified, and include minerals formed by chemical alteration of the primary minerals, or by the deposition of new minerals in an igneous rock. *Secondary minerals are not important in the classification of igneous rocks.*

Primary minerals consist of two types for the purpose of classifying igneous rocks—*essential minerals* and *accessory minerals.* The essential minerals are those that must be present in order for the rock to be assigned a specific position in the classification scheme (i.e., given a specific name). Accessory minerals are those that may or may not be present in a rock of a given type, but the presence of an accessory mineral in a given rock may affect the name of the rock. For example, the essential minerals of a granite are quartz and K-feldspar (K-feldspar is a common notation for the group of potassium-rich feldspars of which orthoclase is the most common). If a particular granite contains an accessory mineral such as biotite or hornblende, the rock may be called a biotite granite or a hornblende granite, respectively.

The essential minerals contained in the common igneous rocks are quartz, K-feldspar, plagioclase, pyroxene (commonly augite), amphibole (commonly hornblende), and olivine. The key to igneous rock identification is the ability of the observer to recognize the presence or absence of quartz and to distinguish between K-feldspar and plagioclase. Color is of little help in the matter because quartz, K-feldspar, and plagioclase can all occur in the same shade of gray. The distinction between quartz and the feldspars is made by the fact that quartz has no cleavage; macroscopic quartz crystals do not exhibit shiny cleavage faces as the feldspars do.

TABLE 1.7 *Classification and Identification Chart for Hand Specimens of Common Igneous Rocks*

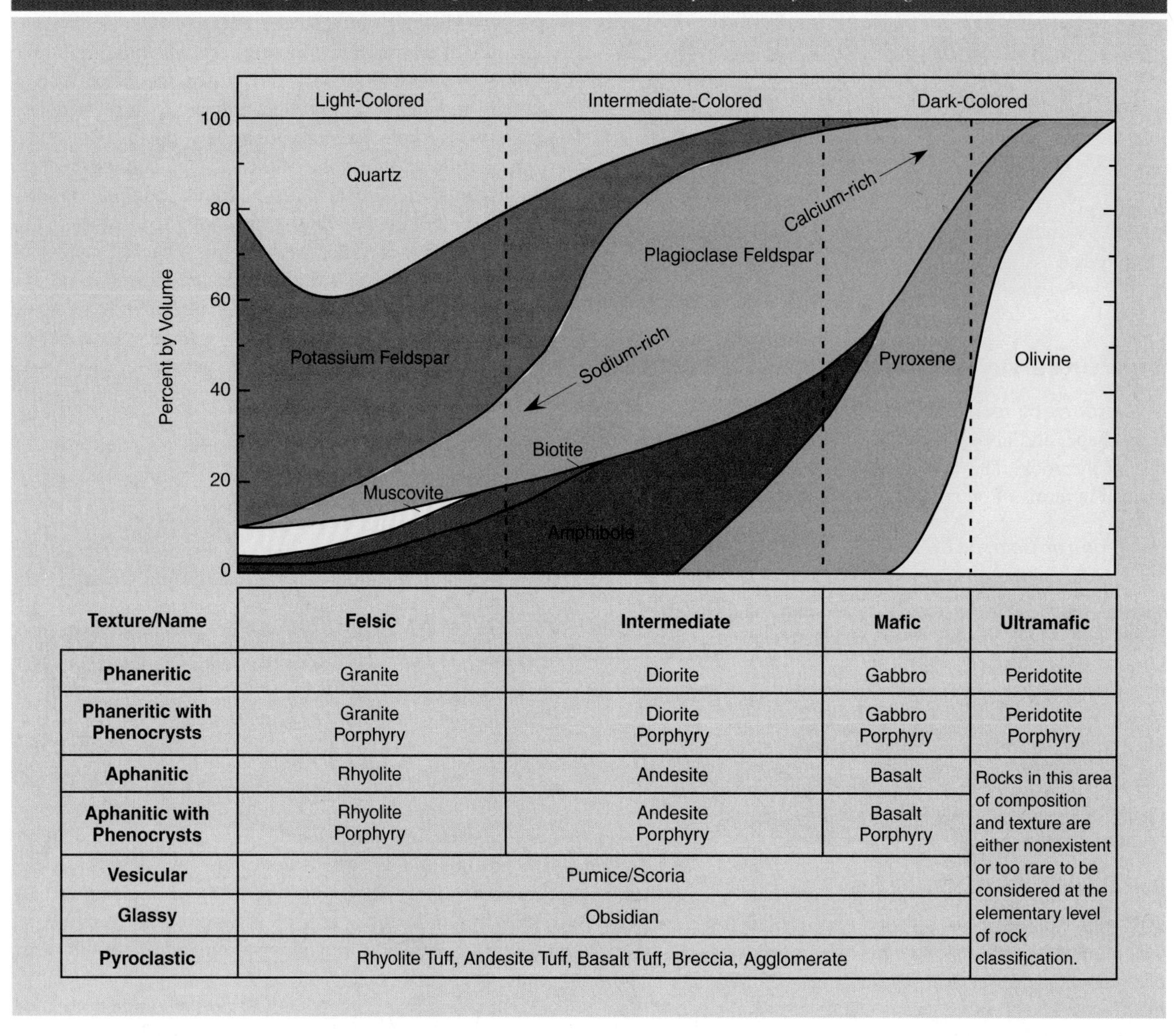

Texture/Name	Felsic	Intermediate	Mafic	Ultramafic
Phaneritic	Granite	Diorite	Gabbro	Peridotite
Phaneritic with Phenocrysts	Granite Porphyry	Diorite Porphyry	Gabbro Porphyry	Peridotite Porphyry
Aphanitic	Rhyolite	Andesite	Basalt	Rocks in this area of composition and texture are either nonexistent or too rare to be considered at the elementary level of rock classification.
Aphanitic with Phenocrysts	Rhyolite Porphyry	Andesite Porphyry	Basalt Porphyry	
Vesicular	Pumice/Scoria			
Glassy	Obsidian			
Pyroclastic	Rhyolite Tuff, Andesite Tuff, Basalt Tuff, Breccia, Agglomerate			

The distinction between K-feldspar and plagioclase is more difficult because both have cleavage faces that show up as shiny surfaces in phaneritic hand specimens. A pink-colored feldspar is usually K-feldspar (orthoclase), but a white or gray feldspar may be either K-feldspar or plagioclase. Plagioclase, however, has characteristic striations that may be visible on cleavage faces, especially if the crystals are several millimeters in size. Figure 1.10 shows a large fragment of a plagioclase crystal with striations on it. Striations on smaller plagioclase crystals are difficult but not impossible to detect.

Colors of Igneous Rocks

The mineral constituents of an igneous rock impart a characteristic color to it. Hence, rock color is used as a first-order approximation in establishing the general mineralogic composition of an igneous rock. As already pointed out, color is a relative and subjective property when modified only by the adjectives *light, intermediate,* or *dark.* All observers would agree that a white rock is "light-colored," a black rock is "dark-colored," and a rock with half of its constituent minerals white and half of them black is a rock of "intermediate color." Mineral constituents are not all black or white, however. Some are pink, gray, and other colors, a fact that adds to the difficulty encountered in igneous rock classification for the beginning student. Nevertheless, the terms *light, intermediate,* and *dark* are useful terms, especially in the classification of hand specimens with an aphanitic texture. For example, the rocks in figures 1.25, 1.26, and 1.33 are light-colored. The rock in figure 1.29 is intermediate in color, and the rocks in figures 1.28 and 1.30 are dark-colored.

Igneous Rock Classification

Table 1.7 is a chart that shows the names of about twenty common igneous rocks and their corresponding textures and mineralogic compositions. This classification scheme is a simplified version of a more complex classification system in which the number of rock names is three to four times the number contained in table 1.7. Some simplification at the level of the beginning student is a pedagogical necessity, and if you should find a rock in your laboratory collection that does not fit easily into one of the categories shown in the table, it may be because the table has been so simplified. For example, *pegmatite* (see figure 1.27) is an example of a very coarse-grained igneous rock commonly associated with the margins of plutons. In general the mineralogic composition of pegmatites is granitic, but they may contain extremely large crystals of uncommon minerals such as spodumene, beryl, tourmaline, or topaz. This type of rock texture does not easily fit into a simplified rock classification table and thus has not been included in table 1.7.

In table 1.7 the lower left-hand side of the chart presents the textural categories while the color categories are displayed across the top. The upper part of the chart shows the mineral constituents of the various rock types. The width of the bar for each mineral is roughly proportional to the percentage of that particular mineral present. For example, quartz ranges from 0% to 25% as a mineral constituent, potassium feldspar (orthoclase) from 0% to 80%, plagioclase from 10% to 70%, olivine from 0% to 90%, amphibole from 0% to 25%, and pryoxene from 0% to 30%. The highest percentage value is equivalent to the widest part of the bar. These bars are used as estimates of percentages for the various mineral constituents and not as rigorous constraints. This is in keeping with the fact that only with phenocrysts or in phaneritic rocks can the percentage of minerals in a given hand specimen be approximated by visual inspection.

The classification of aphanitic, vesicular, and glassy rocks by macroscopic means is dependent on color alone. For this reason, the chart contains alternate names that can be used if a particular specimen does not fit neatly into one of the boxes on the chart. For example, an aphanitic rock that seems too dark to be a rhyolite but too light to be an andesite can be called a *felsite.* Similarly, a rock intermediate between a granite and a diorite can be labeled a *granitic rock.*

A group of igneous rocks that do not fit easily into the general classification scheme are the *pyroclastic rocks,* those that are accumulations of the material ejected from explosive-type volcanoes. The lavas of these volcanoes are characterized by *high viscosity* (they do not flow easily) and high silica content. They are rhyolitic or andesitic in composition. Their mineral constituents may be difficult to determine.

The volcanic ash generated from an eruption is known as *tuff* when it becomes consolidated into a rock. Light-colored tuff is called *rhyolite tuff* (fig. 1.35), and tuff of intermediate color is called *andesite tuff.* In some tuffs, small beadlike fragments of volcanic glass occur. These features are called *lapilli,* and the rock containing them is *lapilli tuff.*

A rock composed of the angular fragments from a volcanic eruption is a *volcanic breccia.* A volcanic rock composed of volcanic bombs and other rounded fragments is known as an *agglomerate.*

Name *Section* *Date*

EXERCISE 2

Identification of Common Igneous Rocks

The identification and classification of igneous rocks is based on their texture and mineralogy. Color is also useful, but it usually is a reflection of the mineral composition of the rock.

A collection of igneous rock hand specimens will be provided to you by your instructor. Figures 1.25 through 1.40 show some examples of hand specimens of common igneous rocks. Although these may be helpful in identification, remember that the color of igneous rocks may vary considerably. For example, compare figures 1.25 and 1.26. Recognition of texture and mineralogy are the keys to the identification of igneous rocks in hand specimens.

Using the worksheets provided on the following pages to record your observations, follow the steps outlined below.

1. Group the specimens into the textural categories described for you in the text. Note the variation in grain size for the phaneritic (coarse-grained) specimens. Separate out those specimens with diagnostic features such as phenocrysts or vesicles.
2. For the phaneritic (coarse-grained) specimens, identify the minerals present in each specimen.
3. Using table 1.7, proceed to identify and name each specimen in your collection. Note that you may have more than one specimen of a given rock type. If you identify several specimens as granite, try to identify accessory minerals in each so that the name you assign is more definitive than just "granite." For example, a granite containing biotite should be called a *biotite granite.*
4. Your laboratory instructor will advise you as to the procedure to be used to verify your identification.

FIGURE 1.25 Pink granite.

FIGURE 1.26 White granite.

FIGURE 1.27 Pegmatite.

FIGURE 1.28 Gabbro.

FIGURE 1.29 Diorite.

FIGURE 1.30 Basalt.

FIGURE 1.31 Vesicular basalt.

FIGURE 1.32 Basalt porphyry.

FIGURE 1.33 Rhyolite.

FIGURE 1.34 Rhyolite porphyry.

FIGURE 1.35 Rhyolite tuff.

FIGURE 1.36 Andesite porphyry.

FIGURE 1.37 Pumice.

FIGURE 1.38 Scoria.

FIGURE 1.39 Obsidian.

FIGURE 1.40 Peridotite.

WEB CONNECTIONS

ROCKS

http://www.geolab.unc.edu/Petunia/IgMetAtlas/mainmenu.html

Atlas of igneous and metamorphic rocks, minerals, and textures. Part of the "Virtual Geology" project at the University of North Carolina. Provides an index of minerals and images of plutonic microtextures, volcanic microtextures, and metamorphic microtextures.

http://www.science.ubc.ca/~geol202/

This is a basic petrology course on the Web. The majority of the material may be too advanced for beginning students but some may find the materials presented useful and interesting. Igneous, sedimentary, and metamorphic rocks are discussed.

GENERAL INTEREST

http://www.rahul.net/infodyn/rockhounds/rockhounds.html

A site on the Web providing information to the "Rockhounds" of the world. Mailing lists, shops and galleries, images and pictures, etc.; links to all are provided. Primarily a commercial site but lots of good information.

Sedimentary Rocks

Origin

Sedimentary rocks are derived from preexisting materials through the work of mechanical or chemical agencies under conditions normal at the surface of the earth, or they may be composed of accumulations of organic debris. Rock weathering on land produces rock and mineral *detritus* (fragments) that are transported by gravity (falling) or by wind, water, or ice and deposited elsewhere on the earth's surface as *clastic sediments.* Weathering also dissolves rock material and makes it available in solution to streams, rivers, and groundwater that transport it to lakes and oceans where it may be deposited as a chemical precipitate or an evaporite.

After sediment has been deposited it may be compacted and cemented into a coherent mass or sedimentary rock. Common cements are silica, calcium carbonate, and iron oxide. This process of changing the soft sediment into rock is known as *lithification,* and it may involve a number of changes resulting from heat, pressure, biological activity, and reaction with circulating groundwaters carrying other materials in solution. The sum of these changes is called *diagenesis* by geologists. Sedimentary rocks display various degrees of lithification. In this manual we will be concerned only with those sedimentary rocks that are sufficiently lithified to permit their being displayed and handled as coherent hand specimens.

Sedimentary rocks provide a number of clues as to their history of transportation and deposition. Some of these features may be present in the hand specimens that will be available in the laboratory. These *sedimentary structures* include such features as bedding, cross-bedding, graded bedding, ripple marks, and mud cracks, among others. In addition, some rock types are characteristic of specific environments of deposition.

Sedimentary rock classification is based on texture and mineralogic composition. Both features are related to the origin and lithification of the original sediment, but the origin cannot be inferred from a single hand specimen. Therefore, the classification scheme used will emphasize the physical features and mineralogy of the rock rather than its exact mode of origin.

Occurrence of Sedimentary Rocks

Sedimentary rocks are formed in a wide variety of sedimentary environments. Any places that sediments accumulate are sites of future sedimentary rocks. *Sedimentary environments* are grouped into three major categories: *continental, mixed (continental and marine),* and *marine.* Within each of these broad categories several subcategories exist. An abbreviated classification of sedimentary environments is given in table 1.8. This table is not intended to show all of the possible environments of deposition but rather to illustrate the environments in which most of the sedimentary rock types you will study here might have originated.

Because the earth's surface is dynamic and ever-changing, sedimentary environments in a given geographic location do not remain constant throughout geologic time. Areas that are now above sea level may have been covered by the sea at various times during the geologic past. Glaciers that existed in the geologic past have since disappeared. Lakes have formed or have dried up. As the depositional environment in a given locality changes through time, the kind of sediment deposited there changes in response to the new environmental conditions.

Sediments that accumulate in a particular sedimentary environment are generally deposited in layers that are horizontal or almost horizontal. The lateral continuity of these layers, or *beds,* reflects the areal extent and uniformity of the environment in which they were deposited. The thickness of a particular bed is a function of the *time* during which the depositional environment remained more or less constant and the *rate* at which the sediment accumulated. Separating the beds are *bedding planes,* which represent a

TABLE 1.8 *Simplified and Abbreviated Classification of Sedimentary Environments and Some of the Rock Types Produced in Each*

SEDIMENTARY ENVIRONMENT	SEDIMENTARY ROCK TYPE
CONTINENTAL	
Desert	Sandstone
Glacial	Tillite
River Beds	Sandstone, Conglomerate
River Floodplains	Siltstone
Alluvial Fans	Arkose, Conglomerate, Sandstone
Lakes	Shale, Siltstone
Swamps	Lignite, Coal
Caves, Hot Springs	Travertine
MIXED (CONTINENTAL AND MARINE)	
Littoral (between high and low tide)	Sandstone, Coquina, Conglomerate
Deltaic	Sandstone, Siltstone, Shale
MARINE	
Neritic (low tide to edge of shelf)	Sandstone, Arkose, Reef limestone, Calcarenite
Bathyal (400 to 4000 m depth)	Chalk, Rock salt, Rock gypsum, Shale, Limestone, Graywacke
Abyssal (depths more than 4000 m)	Diatomite, Shale

TABLE 1.9 *Simplified Particle Size Range, Clastic Sediment Name, and Associated Sedimentary Rock Type*

PARTICLE SIZE RANGE	SEDIMENT	ROCK
Over 256 mm (10 in)	Boulder	Conglomerate (rounded fragments) or breccia (angular fragments)
2 to 256 mm (0.08 to 10 in)	Gravel	Conglomerate or breccia
1/16 to 2 mm (0.025 to 0.08 in)	Sand	Sandstone
1/256 to 1/16 mm (0.00025 to 0.025 in)	Silt	Siltstone*
Less than 1/256 mm (less than 0.00015 in)	Clay	Shale (claystone*)

*Both siltstone and claystone are also known as mudstone, commonly called *shale* if the rock shows a tendency to split on parallel planes. (From Carla W. Montgomery, *Fundamentals of Geology,* 2d ed. Copyright © 1993 McGraw-Hill Company, Inc., Dubuque, Iowa. All rights reserved. Reprinted by permission.)

slight or major change in the depositional history. These horizontal surfaces tend to be surfaces along which the sedimentary rocks break or separate. Rates of deposition vary widely, and a layer of shale 10 feet thick may represent a longer period of geologic time than a sandstone layer 100 feet thick.

When the environment of deposition changes in a particular region, the nature of the sediments accumulating there also changes. A bed of sandstone on top of a bed of shale reflects a change in sedimentary conditions from one in which clay was deposited to one in which sand was deposited. Both the underlying shale and the overlying sandstone represent the passage of geologic time.

TEXTURES OF SEDIMENTARY ROCKS

Texture is the appearance of a rock that results from the size, shape, and arrangement of the mineral grains in the rock. The texture of sedimentary rocks can be described in terms of one of the following: clastic, crystalline, amorphous, oolitic, or bioclastic.

Clastic

Rocks with a *clastic* texture are derived from the detritus or rock waste that has been transported and deposited. (NOTE: In some cases, the terms *detrital or detritus* are used synonymously with clastic as a textural term. We have chosen to use detritus as the source, clastic as the resulting texture.)

The size of the individual particles is one of the chief means of distinguishing sedimentary rock types, and clastic particles are named according to their dimensions (i.e., their average grain diameters). Table 1.9 shows the names and sizes of sedimentary particles that will be useful in the classification of sedimentary rocks. This is a simplified version of a much more detailed classification system.

Particles larger than about ¼ mm can be distinguished with the naked or corrected eye. (Grains of ordinary table salt range in size from about ¼ to ½ mm in diameter. The dot in the letter "i" is about ½ mm in diameter, and a lowercase "l" is about 2½ mm high.) Sand grains smaller than ¼ mm and the larger silt grains can be distinguished with a hand lens, but the smaller grains of silt and clay can be distinguished only with a microscope. Be aware that the term *clay* is used here to define grain size rather than mineralogy, and take care to use the terms *"clay mineral"* and *"clay size"* to avoid confusion.

Sedimentary rocks composed mainly of silt or clay-size particles are said to have a *dense* texture. The term *dense* is also a general textural term for any rock in which the individual mineral components are microscopic in size.

The shapes of the particles in the rock also influence texture. The mineral grains or rock fragments in a specimen with a clastic texture should be examined to determine whether they are *angular* or *rounded.*

Clastic texture is also influenced by the variation of particle size in the rock. Rocks with a predominant grain size are said to be *well sorted;* those with a wide variation in grain size are said to be *poorly sorted.*

Crystalline

This texture is characteristic of a sedimentary rock composed of interlocking crystals. If the individual crystals are less than ¼ mm in diameter, the rock has a *dense texture* as far as macroscopic examination of a hand specimen is concerned.

Amorphous

This is a very dense texture found in rocks composed of finely divided noncrystalline material deposited by chemical precipitation.

Oolitic

This texture is formed by spheroidal particles less than 2 mm in diameter, called *oolites.* The oolites are usually composed of calcium carbonate or silica. They form by deposition of the material out of solution onto a nucleus (much like the formation of a hailstone), and they are cemented together into a coherent rock.

Bioclastic

This texture is produced by the aggregation of fragments of organic remains, the most common of which are shell fragments or plant fragments. Rocks that contain fragments, molds, or casts of organic remains or *fossils* (e.g., shells, bones, teeth, leaves, seeds) or other recognizable evidence of past life (e.g., footprints, leafprints, worm burrows, etc.) are said to be *fossiliferous.* Fossils are commonly embedded in a matrix of sandstone, shale, limestone, or dolomite.

SEDIMENTARY STRUCTURES

Sedimentary rocks may contain features that are characteristic of their environment of deposition. These include *mud cracks, graded bedding, ripple marks, cross bedding,* and *sole marks* among others. Even in hand specimens, many of these features may be evident and provide some indication of the sedimentary environment in which the rock was formed. For example, the *mud cracks* found in rocks formed in exactly the same way as the mud cracks we see formed on dry mud today (fig. 1.41*A*). *Graded bedding* (fig. 1.41*B*) is the gradual vertical shift from coarse to fine clastic material in the same bed.

Oscillation (symmetric) ripple marks (fig. 1.42*A*) are characteristic of sediments deposited where there was a forward and backward movement of water such as one might find in a standing body of water affected by wave action. *Current (asymmetric) ripple marks* (fig. 1.42*B*) indicate that the sediment was deposited by running water or by wind. *Cross-bedding* is characteristically laid down at an angle to the horizontal, such as on the lee side of a sand dune, and while the major bedding is horizontal, there is a subset that is at an angle (fig. 1.42*C*). Sole marks occur when there has been some disturbance of the top of a soft sediment that is then preserved when additional material is deposited. These markings are often tracks and trails of bottom-dwelling organisms.

A.

B.

FIGURE 1.41

A. Mud cracks forming in modern clay-rich sediment (plan view).
B. Schematic cross section of the formation of graded bedding.

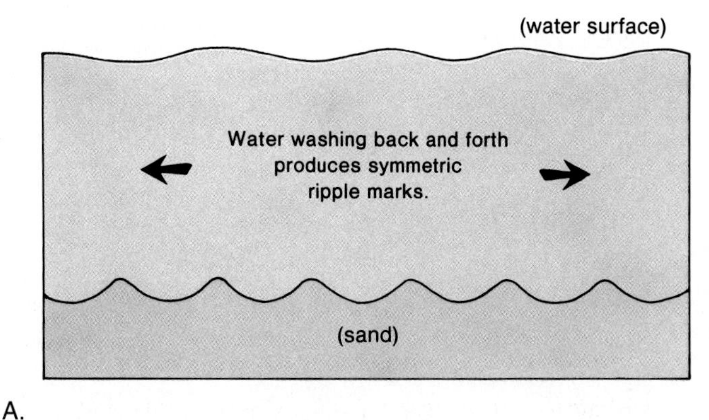

A.

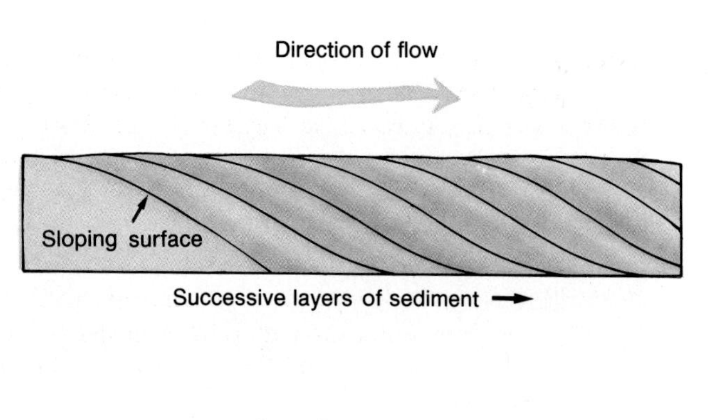

B.

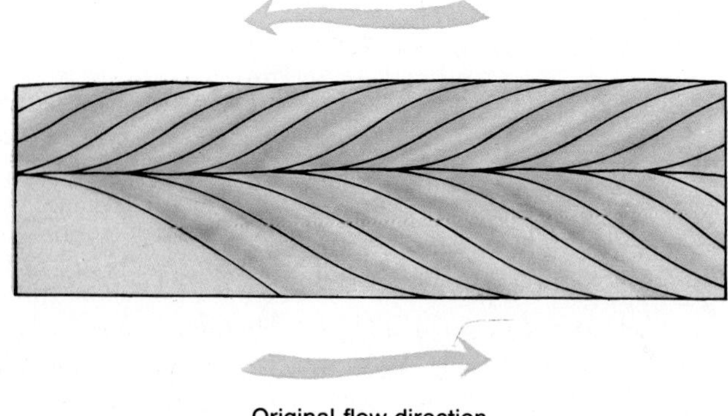

C.

FIGURE 1.42

A. Schematic of the formation of oscillation ripple marks.
B. Schematic of the formation of current ripple marks.
C. Schematic example of cross-bedding formation.

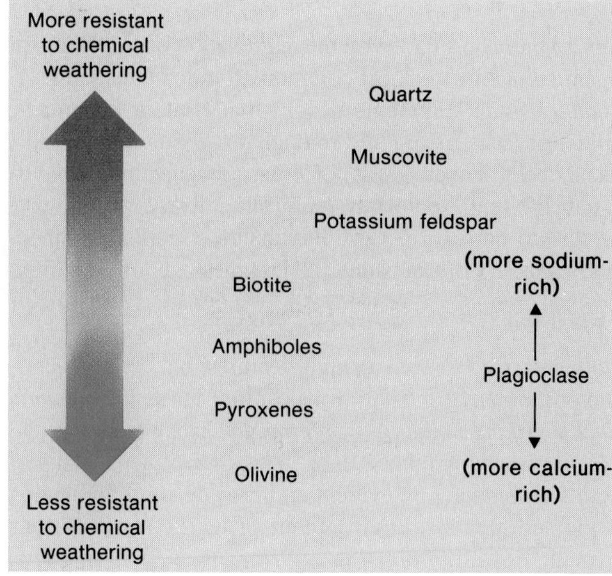

FIGURE 1.43

Susceptibility of minerals to chemical weathering is inversely related to Bowen's reaction series.

COMPOSITION OF SEDIMENTARY ROCKS

In hand specimen, the mineral constituents of sedimentary rocks are generally less varied than those in igneous rocks. The weathering of preexisting rocks involves chemical reactions that act upon the minerals within the rocks. Calcite, for example, may be dissolved and go into solution. In other minerals, these reactions result in a complete alteration of the mineral and often result in new minerals.

The stability of minerals at the surface of the earth varies greatly. Recall the reaction series discussed in the section on igneous rocks (see fig. 1.22). During the weathering process, those minerals that formed at high temperatures are the least stable at the surface of the earth and are most susceptible to chemical weathering. Figure 1.43 relates the Bowen's reaction series to mineral stability. Thus, it is not surprising that olivine grains are relatively rare in sedimentary rocks and that quartz is common. The end products of the chemical weathering of most silicate minerals are quartz and clay minerals. A great many chemical reactions occur during the weathering, transportation, deposition, and lithification of sediments, all of which have some influence on the final composition of the rock.

Because some of the minerals that occur are often microscopic in size, they cannot always be identified easily. Tests for hardness and chemical composition are employed where visual inspection alone fails to identify the primary substance of which a sedimentary rock is composed.

Discussion of materials that are common in sedimentary rocks follows. Some are shown in the suite of sedimentary rocks in figures 1.44 through 1.51.

Silica

Silica, SiO_2, commonly occurs as grains in quartz sandstone. Quartz is the most ubiquitous of all minerals in sedimentary rocks. Other forms of silica include *chert,* a dense, cryptocrystalline form of quartz, and *diatomite,* a porous accumulation of the remains of siliceous plants of microscopic size.

Carbonates

Calcite, $CaCO_3$, and dolomite, $CaMg(CO_3)_2$, are two common minerals that occur as the major constituents of limestone (calcite) and dolomite (or dolostone to distinguish it from the mineral), or as the cementing material in a wide variety of clastic sediments. A sedimentary rock containing calcite in any form will effervesce (fizz) strongly when a drop of cold dilute HCl is placed on it. Powdered dolomite will react weakly with the same cold dilute HCl. When applying this test it is important to distinguish whether it is the grains or crystals that are reacting with the acid, or whether it is the cement that is a carbonate.

The names of sedimentary rocks in which a carbonate mineral is present but is not a major constituent are prefixed by the term *calcareous,* which designates the presence of significant amounts of calcite or dolomite. For example, a rock made up of sand-sized quartz grains cemented with calcite would be called a *calcareous quartz sandstone.*

Calcite occurs in the crystalline form in *crystalline limestone.* Many shell fragments are composed of calcite, and the resulting rocks may range from *fossiliferous limestone* (fig. 1.49) to rocks called *coquina* composed of almost 100% shell fragments (fig. 1.51). Most oolites are composed of calcite. Thus, there is a wide range of rocks called limestone ranging from the crystalline variety to coquina.

Clay Minerals

This term refers to a group of *silicate* minerals that have layered atomic structures. *Clay minerals* are not to be confused with *clay-size particles,* which are smaller than 1/256 mm in diameter. All clay minerals occur as clay-size particles, but not all clay-size particles are clay minerals.

Although one clay mineral, kaolinite, is white, most clay minerals are green to gray and impart a dark color to the sedimentary rock in which they occur. Clay minerals are most common in dense fine-grained rocks such as *mudstone, shale* (distinguished by its *fissility,* the tendency to part in thin layers), and *graywacke.* Many limestones contain appreciable amounts of clay. The adjective used to describe a rock containing some clay is *argillaceous,* as for example an *argillaceous limestone.*

Evaporites

Minerals that belong to this group include gypsum, $CaSO_4{\cdot}2H_2O$, and halite, NaCl. Both are formed by chemical precipitation from an aqueous solution. Gypsum may occur as crystals in sedimentary rocks, but in this form it usually formed after the rock was deposited.

Rock Fragments

Some sedimentary rocks contain very coarse detrital constituents such as pebbles, cobbles, or even boulders. These coarse materials are usually rock fragments rather than single minerals. *Conglomerates* are clastic sedimentary rocks containing rounded pebbles or cobbles (fig. 1.46). If the coarse rock fragments are angular, the rock name is sedimentary *breccia* (fig. 1.48).

Feldspars

Compared to quartz, calcite, and clay minerals, feldspars do not occur in great abundance in sedimentary rocks. Under certain circumstances, however, some sedimentary rocks may contain significant amounts of feldspar. A feldspar-rich sandstone is called *arkose* (fig. 1.45). The corresponding adjective for rocks with some feldspar is *arkosic.* Feldspar also occurs in variable amounts in graywacke.

Heavy Minerals

Heavy minerals such as garnet, hornblende, ilmenite, magnetite, rutile, tourmaline, and zircon, among others, may be found in sedimentary rocks but they rarely exceed one percent of the total volume. These minerals may be useful in determining the source of the sediments that make up the rock.

Organic Constituents

As noted above, organic remains may make up significant amounts of a given sedimentary rock. These materials may include shell fragments, teeth, bones, plant remains, or other organic debris.

Classification of Sedimentary Rocks

The great variety of sedimentary environments in nature and the gradations between them are responsible for the large diversity of sedimentary rock types. Because of this, a classification scheme encompassing all possible varieties would be unduly complex for the beginning student.

The threefold classification of sedimentary rocks presented in table 1.10 is highly simplified and is based on the major constituents of sedimentary rocks: *inorganic detrital materials, organic detrital materials,* and *inorganic chemical precipitates.* This arrangement is a logical and useful guide for an understanding of the similarities and differences in sedimentary rock types encountered by the beginning student.

The identification scheme for use with table 1.10 is presented as a part of Exercise 3.

FIGURE 1.44 Sandstone.

FIGURE 1.45 Arkose.

FIGURE 1.46 Conglomerate.

FIGURE 1.47 Shale.

FIGURE 1.48 Breccia.

FIGURE 1.49 Fossiliferous limestone.

FIGURE 1.50 Chalk.

FIGURE 1.51 Coquina.

TABLE 1.10 *Classification and Identification Chart for Hand Specimens of Common Sedimentary Rocks*

Origin	Textural Features and Particle Size	Composition and/or Diagnostic Features	Rock Name
INORGANIC DETRITAL MATERIALS	Clastic. Pebbles and granules embedded in matrix of cemented sand grains.	Angular rock or mineral fragments.	SEDIMENTARY BRECCIA
		Rounded rock or mineral fragments.	CONGLOMERATE
	Clastic. Coarse sand and granules.	Angular fragments of feldspar mixed with quartz and other mineral grains. Pink feldspar common.	ARKOSE
	Clastic. Sand-size particles.	Rounded to subrounded quartz grains. Color: white, buff, pink, brown, tan.	QUARTZ SANDSTONE
		Calcite and/or dolomite grains. Light-colored.	CALCARENITE
	Clastic. Sand-size particles mixed with clay-size particles.	Quartz and other mineral grains mixed with clay. Color: dark gray to gray-green.	GRAYWACKE
	Clastic. Fine-grained. Silt and clay-size particles.	Mineral constituents (commonly quartz) may be identifiable with hand lens. Usually well stratified. Color: varies.	MUDSTONE: SILTSTONE
		Mineral constituents not identifiable. Soft enough to be scratched with fingernail. Usually well stratified. Fissile (tendency to separate in thin layers). Color: varies.	MUDSTONE: SHALE
		Mineral constituents not identifiable. Soft enough to be scratched with fingernail. Massive (earthy). Color: varies.	MUDSTONE: CLAYSTONE
INORGANIC CHEMICAL PRECIPITATES	Dense, crystalline or oolitic.	$CaCO_3$; effervesces freely with cold dilute HCl. May contain fossils (*fossiliferous*). Some varieties are *crystalline*. Some varieties are *oolitic*. Color: white, gray, black. Generally lacks stratification.	LIMESTONE
	Dense or crystalline.	$CaMg(CO_3)_2$; powder effervesces weakly with cold dilute HCl. May contain fossils (*fossiliferous*). Color: varies, but commonly similar to limestones. Stratification generally absent in hand specimens.	DOLOMITE
	Dense, porous.	$CaCO_3$; effervesces freely with cold dilute HCl. Color: varies. Contains irregular dark bands.	TRAVERTINE
	Dense (amorphous).	Scratches glass, conchoidal fracture. Color: black, white, gray.	CHERT
	Crystalline.	$CaSO_4 \cdot 2H_2O$; commonly can be scratched with fingernail. Color: varies; commonly pink, buff, white.	ROCK GYPSUM
		NaCl. Salty taste. Color: white to gray. Crystalline. May contain fine-grained impurities in bands or thin layers.	ROCK SALT
ORGANIC DETRITAL MATERIALS	Earthy (bioclastic).	$CaCO_3$; effervesces freely with cold dilute HCl; easily scratched with fingernail. Microscopic organisms. Color: white.	CHALK
		Soft. Resembles chalk but does not react with HCl. Commonly stratified. Color: gray to white. Microscopic siliceous plant remains.	DIATOMITE
	Bioclastic.	$CaCO_3$; calcareous shell fragments in a massive or crystalline matrix. Effervesces freely with cold dilute HCl.	LIMESTONE
	Bioclastic.	$CaCO_3$; calcareous shell fragments cemented together.	COQUINA
	Fibrous (bioclastic).	Brown plant fibers. Soft, porous, low specific gravity.	PEAT
	Dense (bioclastic).	Brownish to brown-black. Harder than peat.	LIGNITE
	Dense (bioclastic).	Black, dull luster. Smudges fingers when handled.	BITUMINOUS COAL

Name Section Date

EXERCISE 3

IDENTIFICATION OF COMMON SEDIMENTARY ROCKS

A collection of sedimentary rocks will be provided in the laboratory by your instructor. Take time to examine the specimens so that you have some familiarity with the various types provided. In figures 1.44 through 1.51, examples of some sedimentary rocks are shown. Remember that there is a wide diversity of sedimentary rocks, and these examples may not exactly represent the rock types provided to you. As you complete the following parts of the exercise, keep a written record of the procedures you followed to determine the name for each specimen.

Using the worksheets provided on the following pages to record your observations, follow the steps outlined below.

1. For each specimen, determine if (A) it is composed of mineral or rock fragments (inorganic detrital materials), or (B) it is generally light-colored and crystalline, oolitic, or dense (chemical precipitates), or (C) it is composed of fossil materials from animals or plants (organic detrital materials).
2. For each specimen in group A, estimate the grain size and shape. Then determine the major constituent (i.e., quartz grains, rock fragments, etc.). Test with a drop of cold dilute HCl if you suspect the presence of carbonate. Using the appropriate part of table 1.10, determine the rock name for each specimen in this group.
3. For those specimens in group B, test with a drop of cold dilute HCl (remembering that dolomite will react with cold dilute HCl only when the dolomite is in a powdered form).
   (a) If there is a reaction, examine the specimen for other diagnostic features such as oolites, possibly some fossil remains, etc. Determine the rock name from table 1.10 for each specimen in this category.
   (b) If there is no reaction, test for hardness. (A taste test may be useful under appropriate circumstances, but as noted on page 7, for obvious sanitary reasons do not use the taste test on laboratory hand specimens.)
4. For those specimens in group C, sort out the light-colored, fine-grained rocks.
   (a) Apply a drop of cold dilute HCl to the light-colored fine-grained specimens. Test for hardness. Examine them for other features. Determine the rock name from table 1.10 for each specimen in this category.
   (b) For the remaining specimens in this category, the composition and other diagnostic features should allow you to determine the rock name from table 1.10.
5. Where possible, apply the appropriate adjective to the sample; for example, *fossiliferous* limestone or *calcareous* quartz sandstone. Be as specific as possible. This will require special attention to *all* the features of the specimens.
6. Your laboratory instructor will advise you as to the procedure to be used to verify your identification.

METAMORPHIC ROCKS

Once formed, all rocks are subject to processes of change that occur at the surface of the earth or within the crust of the earth. As we have seen, the processes with which we are familiar are those that take place at the surface of the earth and that combine to form sedimentary rocks. *Metamorphic* rocks are formed at varying depths within the crust when preexisting rocks are changed physically or chemically under conditions of high temperature, high pressure, or both. The process of *metamorphism,* literally a "change in form," takes place deep beneath the earth's surface and acts on all rocks—igneous, sedimentary, and metamorphic (refer to the rock cycle diagram, fig. 1.21).

As noted in the discussion of igneous rocks, the geothermal gradient is one source of heat for metamorphism. The other source is the heat from igneous plutons as they rise upward into country rock. Although extremely high temperatures may occur, metamorphism always occurs *below* the melting point of the rocks. If melting occurs, the process is considered igneous.

Pressure derives from two sources. The first is the *confining pressure* on the deeply buried rocks resulting from the weight of the overlying rocks. Confining pressure is equal in all directions and is related to the depth of burial. The second is *directed pressure,* which as the name suggests is greater in one particular direction. An example of directed pressure is that associated with the process of mountain building.

The process of metamorphism is aided by the presence of fluids in the rocks. While it is possible for chemical elements to migrate through the country rock without fluids being present, their movement is greatly facilitated by fluids.

The mineralogy and texture of metamorphic rocks provide some insight into the temperature and pressure environment in which they were formed. *Metamorphic facies* are defined by the typical assemblage of minerals found in the rocks formed at specific combinations of temperature and pressure. *Metamorphic grade* refers to the intensity of metamorphism in a given rock.

METAMORPHIC ENVIRONMENTS

The metamorphism associated with the intrusion of a magma into the country rock (sometimes referred to as the host rocks) results in *contact metamorphism.* Both the heat and the chemical constituents which emanate from the magma produce mineralogical changes in the host rocks. Contact metamorphism is most intense at or near the zone of contact between the magma and the host rock, and the effects of contact metamorphism progressively diminish in a direction away from the contact zone. This *aureole* or halo of metamorphism surrounding the magma is characterized by the formation of high-temperature minerals close to the contact zone and progressively lower temperature minerals as the distance from the contact zone increases. Normally there is no evidence of the reorientation of the minerals. The effect of migration of materials from the magma into the host rock, *metasomatism,* may or may not occur. The size of the aureole depends on the temperature and size of the pluton, mineralogy of the host rock, and the presence or absence of fluids.

By far the largest volume of metamorphic rocks is produced by those processes associated with *regional metamorphism.* Regional metamorphism is a result of large-scale *tectonic* or *mountain-building* events involving movement and deformation of the earth's crust on a regional scale. During periods of mountain building, large segments of crustal rocks are deformed. These deformed areas occur in zones or belts hundreds of miles wide and thousands of miles long (the Appalachian Mountain belt is an example). Rocks in these belts are subjected to stretching (tensional) and squeezing (compressional) stresses that cause physical and mineralogical changes in the rock. They are subjected to extreme stresses, and rocks at great depth may deform as a plastic rather than as a brittle solid. This accounts for the fact that many rocks deformed under conditions of regional metamorphism have a texture, called *foliation,* that is characterized by a parallel arrangement of platy minerals such as the micas or the common orientation of the long axis of such minerals as hornblende.

A third but less important environment in which metamorphism can occur is that associated with *fault zones* such as the San Andreas fault in California (discussed in Exercise 24). In this environment high pressure and low temperature are present, and the materials commonly formed are composed of broken and distorted fragments of the rocks on either side of the fault zone and minerals that form only at low temperature and high pressure. If coarse-grained, these rocks are known as *fault breccia;* if fine-grained, *mylonite.*

METAMORPHIC FACIES AND GRADE

Contact and regional metamorphism produce differing degrees of change in preexisting rocks. As briefly discussed, the aureole associated with contact metamorphism is defined by a decrease in degree of metamorphism as the distance from the pluton increases. The same is true of regional metamorphism; rocks on the outer margins of a mountain belt that has been subjected to a single period of deformation may be only slightly metamorphosed. The metamorphic rocks in the center of the belt may be so deformed that the texture and mineralogy of the original or parent rock have been obliterated.

The study of metamorphic rocks has led to the identification of *metamorphic facies:* the recognition of an assemblage of minerals in the rocks that formed during

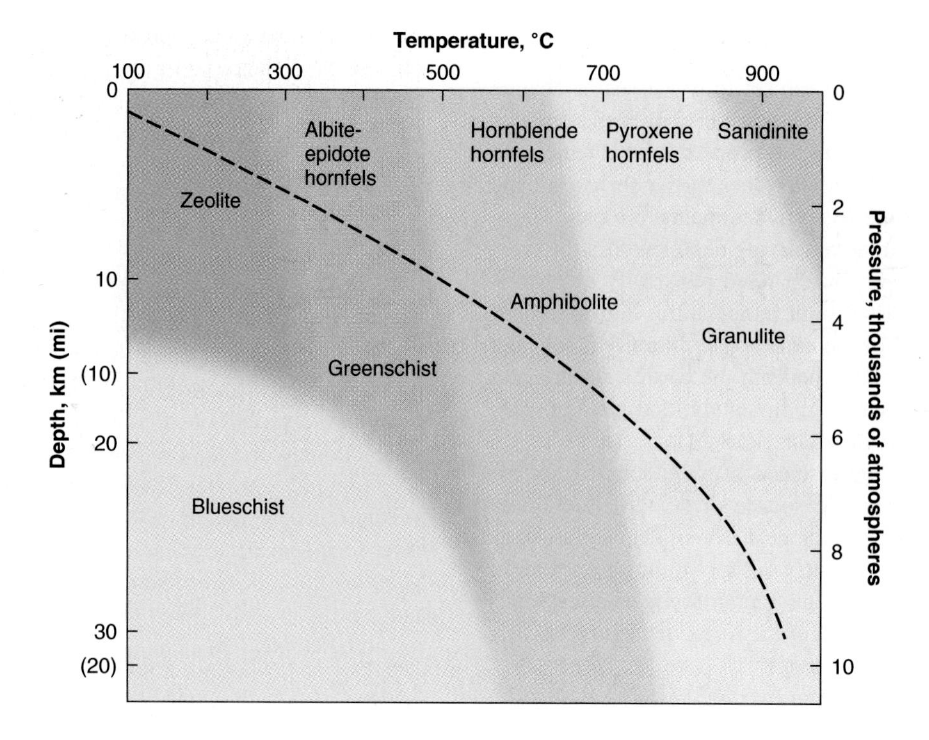

FIGURE 1.52

Metamorphic facies as functions of temperature and pressure (depth). Dashed curve is average continental geothermal gradient.
(From Carla W. Montgomery, Physical Geology, 3d ed. Copyright © 1993 McGraw-Hill Company, Inc., Dubuque, Iowa. *All rights reserved. Reprinted by permission.)*

metamorphism under specific environmental conditions of temperature and pressure. The assemblages formed depend also on the mineralogy of the parent rock, but the same parent rock will yield the same assemblages at the same combination of temperature, pressure, and fluids. The commonly recognized metamorphic facies are named for characteristic minerals or characteristic rock types associated with them (fig. 1.52).

If we consider the effects of contact metamorphism on a single rock type, the resulting facies are relatively simple because temperature is the single important variable. The facies recognized are named for specific minerals: *sanidinite,* the high-temperature facies, is characterized by the presence of the mineral sanidine, a variety of K-feldspar, grading outward from the heat source through the *hornfels* facies to the *zeolite* or low-temperature facies.

The facies relationships associated with regional metamorphism are more complex due to the fact that not only are there two variables, both temperature and pressure, but the size of the area affected is such that there is a wide variety of rock types involved. Not all facies may be present in a given region. Figure 1.52 is a schematic diagram of the temperature and pressure relationships of the commonly recognized metamorphic facies.

Metamorphic grade is the measure of the intensity of metamorphism to which a given rock has been subjected.

TABLE 1.11 *Simplified Table Relating Rock Type, Metamorphic Grade, and Index Minerals*

Rock Type	Metamorphic Grade	Index Mineral
Slate	Low	
Phyllite	Low to intermediate	Chlorite
Schist	Intermediate to high	Garnet
Gneiss	High	Sillimanite
Granulite	Very high	

Low-grade metamorphism occurs in the marginal area; *high-grade* metamorphism occurs where the effects of temperature and pressure have been most intense; and *intermediate-grade* metamorphism lies in between. Table 1.11 is a simplified schematic showing the relationship between metamorphic grade, rock type, and the *index minerals* associated with each. Index minerals are those minerals that are stable over a specific range of temperature and pressure conditions and that are useful in determining metamorphic grade (fig. 1.53).

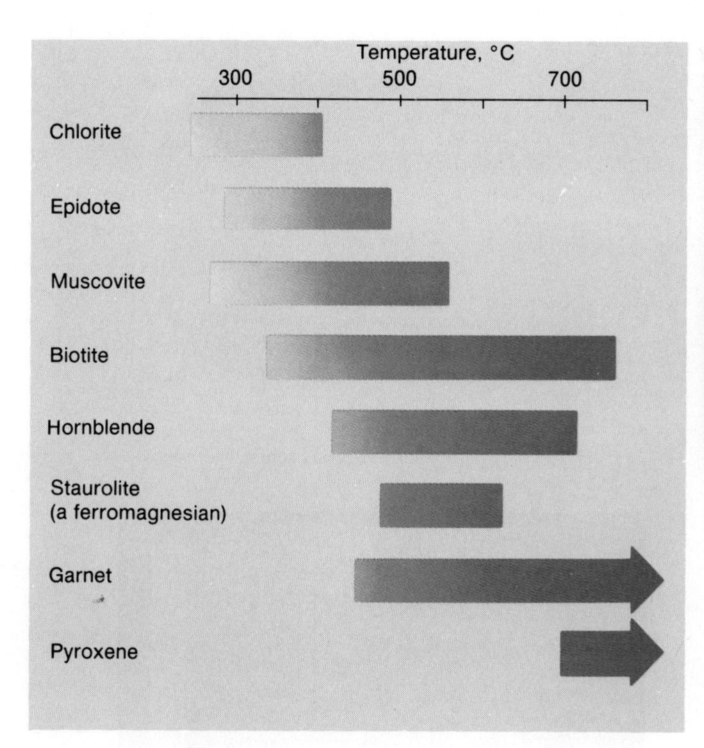

FIGURE 1.53

Approximate temperature ranges over which some representative index minerals are stable.

(From Carla W. Montgomery, Physical Geology, 3d ed. Copyright © 1993 McGraw-Hill Company, Inc., Dubuque, Iowa. *All rights reserved. Reprinted by permission.)*

FIGURE 1.54

Slate, showing characteristic slaty cleavage.

The metamorphism of shale, a common sedimentary rock, provides a simple way of illustrating metamorphic grade. The sequence of increasing metamorphic grade shows the gradual change from shale to slate to phyllite to schist to gneiss to granulite (see table 1.11). In this simplified presentation, slate represents low-grade metamorphism; phyllite, low to intermediate; schist, intermediate to high; and gneiss and granulite, high to very-high grade metamorphism. As metamorphic grade increases, there are corresponding mineralogic and textural changes that occur within the rocks.

Texture and Composition of Metamorphic Rocks

Metamorphic textures consist of two main types, *foliated* and *nonfoliated.* The mineral constituents of foliated metamorphic rocks are oriented in a parallel or subparallel arrangement. Foliated metamorphic rocks are generally associated with regional metamorphism. The nonfoliated rocks exhibit no preferred orientation of mineral grains and are commonly composed of a single mineral.

Foliated Textures

Five kinds of foliated textures are recognized. In order of increasing metamorphic grade, these are *slaty, phyllitic, schistose, gneissic* (pronounced "nice-ick") and *granulite.*

Slaty Texture

This texture is caused by the parallel orientation of microscopic grains. The name for the rock with this texture is *slate* (fig. 1.54), and the rock is characterized by a tendency to separate along parallel planes. This feature is a property known as *slaty cleavage.* (Slaty cleavage or rock cleavage is not to be confused with cleavage in a mineral, which is related to the internal atomic structure of the mineral.)

Phyllitic Texture

This texture is formed by the parallel arrangement of platy minerals, usually micas, that are barely macroscopic (visible to the naked or corrected eye). The parallelism is often wavy, or crenulated. The predominance of micaceous minerals imparts a sheen to the hand specimens. A rock with a phyllitic texture is called a *phyllite* (fig. 1.55).

Schistose Texture

This is a foliated texture resulting from the subparallel to parallel orientation of platy minerals such as chlorite or micas. Other common minerals present are quartz and amphiboles. A schistose texture lies between the parallel platy appearance of phyllite and the distinct banding of gneissic texture. The average grain size of the minerals is generally smaller than in a gneiss (see discussion that follows). A rock with schistose texture is called a *schist* (fig. 1.56).

Gneissic Texture

This is a coarsely foliated texture in which the minerals have been segregated into discontinuous bands, each of which is dominated by one or two minerals. These bands range in thickness from 1 mm to several centimeters. The individual mineral grains are macroscopic and impart a striped appearance to a hand specimen (fig. 1.57). Light-colored bands commonly contain quartz and feldspar, and the dark bands are commonly composed of hornblende and biotite. Accessory minerals are common and are useful in applying specific names to these rocks. A rock with a gneissic texture is called a *gneiss.*

FIGURE 1.55 Phyllite.

FIGURE 1.56 Garnetiferous schist.

FIGURE 1.57 Gneiss.

FIGURE 1.58 Granulite.

FIGURE 1.59 Quartzite.

FIGURE 1.60 Pink marble.

FIGURE 1.61 Metaconglomerate.

FIGURE 1.62 Anthracite coal.

Granulitic Texture

This texture is a medium to coarse, even-grained (mosaic) texture that is almost or completely lacking in water-bearing minerals. Quartz, two feldspars, and almandine garnet and, in one group, pyroxenes are the dominant minerals. Quartzo-feldspathic granulites usually show a distinct foliation of thin lenticular aggregates of quartz grains. In some rocks the foliation is accentuated by discontinuous thin films of biotite. Foliation is virtually lacking in pyroxene granulites. A rock with a granulitic texture is called a *granulite* (fig. 1.58).

Nonfoliated Texture

Metamorphic rocks with no visible preferred orientation of mineral grains have a nonfoliated texture. Nonfoliated rocks commonly contain equidimensional grains of a single mineral such as quartz, calcite, or dolomite. Examples of such rocks are *quartzite* (fig. 1.59), formed from a quartz sandstone, and *marble* (fig. 1.60), formed from a limestone or dolomite (dolostone). Conglomerate that has been metamorphosed may retain the original textural characteristics of the parent rock, including the outlines and colors of the larger grain sizes such as granules and pebbles. However, because metamorphism has caused recrystallization of the matrix, the metamorphosed conglomerate is called *metaconglomerate* (fig. 1.61). In some cases the metamorphism has deformed the shape of the granules or pebbles; in this case the rock is called a *stretch-pebble conglomerate.*

Quartzite and metamorphosed conglomerate can be distinguished from their sedimentary equivalents by the fact that they break *across* the quartz grains, not around them. Marble has a crystalline appearance and generally has larger mineral grains than its sedimentary equivalents.

A fine-grained (dense-textured), nonfoliated rock usually of contact metamorphic origin is *hornfels.* Hornfels has a nondescript appearance because it is usually some medium to dark shade of gray, is lacking in any structural characteristics, and contains few if any recognizable minerals in hand specimen.

The metamorphic equivalent of bituminous coal is *anthracite coal* (fig. 1.62). A sequence similar to that presented for the metamorphism of shale would show the change from peat to lignite to bituminous coal to anthracite coal and, under extreme conditions, to the mineral graphite.

The Naming of Metamorphic Rocks

Foliated metamorphic rocks are named according to their texture (e.g., slate, phyllite, schist, or gneiss). In addition to the root name, the name or names of the dominant or distinctive (but not abundant) minerals may be added as descriptors. In some rocks, mineral crystals have formed that are much larger than the matrix in which they are contained. These are called *porphyroblasts,* and recognition of their mineralogy aids in the detailed description of the rock in which they are contained. For example, a schist with recognizable garnet grains (porphyroblasts) would be called a *garnet schist* (some would use the term "garnetiferous" as the descriptor), and the exact textural description would be "a rock with a porphyroblastic schistose texture." Other examples of rock names with mineral descriptors are: quartz-hornblende gneiss, chlorite schist, biotite-garnet schist, garnet gneiss, amphibole schist, and granite gneiss.

The names of slate are commonly modified by a color name, as in green slate, red slate, or black slate.

In the case of nonfoliated rocks, a color prefix is also commonplace in the naming of quartzite or marble. Such names as white marble, pink marble, or variegated marble (meaning a marble containing streaks of several different colors) are commonplace.

TABLE 1.12 *Classification and Identification Table for Hand Specimens of Common Metamorphic Rocks*

TEXTURE	DIAGNOSTIC FEATURES	ROCK NAME
FOLIATED	Slaty texture (slaty cleavage apparent). Dense, microscopic grains. Color variable; black and dark gray common. Also occurs in green, dark red, and dark purple colors.	SLATE
	Phyllitic texture. Fine-grained to dense. Micaceous minerals are dominant. Has a sparkling appearance.	PHYLLITE
	Schistose texture. Medium to fine-grained. Common minerals are chlorite, biotite, muscovite, garnet, and dark elongate silicate minerals. Feldspars commonly absent. Recognizable minerals used as part of rock name. Porphyroblasts common.	SCHIST
	Gneissic texture. Coarse-grained. Foliation present as macroscopic grains arranged in alternating light and dark bands. Abundant quartz and feldspar in light-colored bands. Dark bands may contain hornblende, augite, garnet, or biotite.	GNEISS
	Granulitic texture, medium to coarse, even-grained. Essentially anhydrous. Foliation present in light-colored quartzo-feldspathic rocks.	GRANULITE
NONFOLIATED	Crystalline. Hard (scratches glass). Breaks across grains as easily as around them. Color variable; white, pink, buff, brown, red, purple.	QUARTZITE
	Dense, dark-colored; various shades of gray, gray-green, to nearly black.	HORNFELS
	Texture of conglomerate but breaks across coarse grains as easily as around them. Granules or pebbles are commonly granitic or jasper, chert, quartz, or quartzite. If pebbles are deformed, called *stretch-pebble conglomerate.*	METACONGLOMERATE
	Crystalline. Composed of calcite or dolomite. Color variable; white, pink, gray, among others. Fossils in some varieties.	MARBLE
	Granulitic texture, medium to coarse, even-grained. Essentially anhydrous. Foliation virtually lacking in pyroxene-plagioclase-bearing rocks.	GRANULITE
	Black, shiny luster. Conchoidal fracture.	ANTHRACITE COAL

Classification of Metamorphic Rocks

The classification of metamorphic rocks is based on texture and composition, the basic distinction being between foliated and nonfoliated textures (table 1.12). *The rock names given are only the root names,* because the chart would become too large and cumbersome if all possible varieties of gneiss, schist, marble, and quartzite were listed. You are encouraged to utilize as many descriptors as applicable in naming metamorphic rocks.

Name Section Date

EXERCISE 4

IDENTIFICATION OF COMMON METAMORPHIC ROCKS

A group of metamorphic rock hand specimens will be provided to you in the laboratory. Take time to review the types of foliated textures described on the previous pages. The rocks depicted in figures 1.55–1.62 are examples of several types of metamorphic rocks and may assist you in identification. Remember that there are a wide variety of types, textures, and colors of metamorphic rocks, and your collection may include specimens that do not appear in these figures.

Using the worksheets provided on the following pages to record your observations, follow the steps outlined below.

1. Divide your hand specimens into the two major textural categories—foliated and nonfoliated.
2. Arrange the foliated specimens in sequence from coarse-grained to fine-grained or dense.
   (a) Work first with the coarse-grained rocks to apply a rock name. Examine each specimen carefully to identify specific minerals that are present either in abundance or as porphyroblasts. Utilize these when applying a specific rock name to the specimens of gneissic or schistose texture. Some examples that may be in your collection are:

   GNEISS:
   Biotite gneiss
   Biotite hornblende gneiss
   Quartz feldspar gneiss
   (Others possible)

   SCHIST:
   Garnet schist
   Muscovite schist
   Biotite schist
   Tourmaline mica schist
   Staurolite schist
   Hornblende schist
   Garnet mica schist
   (Others possible)

   (b) The fine-grained to dense rocks with foliation should be examined on the basis of texture and then color. Descriptors are usually not possible with phyllite, and the slates are generally described by color. In some cases, they may contain enough calcium carbonate to react with cold dilute HCl and would be called "calcareous slate."
   (c) Identify and apply rock names to each of the foliated specimens. For each, give an indication of the metamorphic grade represented.
3. Work next with the nonfoliated specimens. Remember the following:
   (a) Marble is composed mainly of calcite or dolomite, hence it is usually softer than glass and reacts with cold dilute HCl in the same way that limestone or dolomite (dolostone) does.
   (b) Quartzite and metaconglomerate are rich in silica (quartz) and scratch glass.
   (c) Graphite as a rock has the same characteristics as the mineral graphite.
   (d) Identify and apply rock names to all specimens. Where possible use descriptors such as color, fossil content, shape of pebbles, etc.
4. Your laboratory instructor will advise you as to the procedure to be used to verify your identification.

The Geologic Column and Relative Geologic Time

Background

An understanding of the relationships between rock units and their *relative ages* is essential in the unraveling of the geologic history of a given area. The relative ages of sedimentary rocks are determined by the application of two basic geologic principles. The first is the *law of original horizontality,* which states that sediments deposited in water are laid down in strata that are horizontal or nearly horizontal. The second is the *law of superposition,* which states that in any undisturbed sequence of sedimentary rocks, the layer at the bottom of the sequence is older than the layer at the top of the sequence.

Through the application of the principles used to establish the relative ages of sedimentary strata around the world, a geologic time scale for all of earth history has been pieced together. The geologic time scale as used in North America is shown in table 1.13. The table is arranged with the oldest geologic ages at the bottom and the youngest at the top. The *absolute ages* given in table 1.13 have been determined by radiometric age determinations on igneous rocks found within these sedimentary sequences.

All areas of the earth's surface were not sites of deposition throughout all of geologic time. Some areas, especially mountain regions, contained preexisting rocks from which sedimentary materials were produced by natural decay and physical breakdown. These sediments were eroded and transported to sites of deposition by various geologic agents such as running water, glaciers, and wind.

A depositional site may change over geologic time to a site where erosion is taking place. Hence, in any given geographic locality, all of geologic time is not represented by a continuous sequence of strata. Many gaps in the sedimentary record occurred in one locality or another across the face of the earth during the passage of geologic time. Therefore, in order to construct a more complete geologic history of a particular area, fragments of that history recorded in rocks from different localities must be pieced together.

The most thoroughly documented part of the geologic time scale, sometimes referred to as the *geologic column,* comes from the geologic strata deposited during the last 600 million years. This segment of geologic time has been divided into subunits called eras, periods, and epochs. The three major eras in order of decreasing age are the Paleozoic, Mesozoic, and Cenozoic. Each of these is divided into periods, and the periods are further divided into epochs. The names of the eras are based generally on the fossils contained in strata formed during those eras. *A fossil is any evidence of past life such as bones, shells, leaf imprints, and the like.* Paleozoic means "early life," Mesozoic means "middle life," and Cenozoic means "recent life." The names of the periods and epochs are based on strata originally studied in Europe during the eighteenth and nineteenth centuries; hence, the names are chiefly European in origin.

The time units of the geologic column are not of equal duration as can be seen from the absolute ages of the time boundaries between the eras in table 1.13. The geologic column as it existed in the early part of the twentieth century was also subdivided by absolute ages, but these dates lacked precision because they were based on inaccurate assumptions. The absolute ages in table 1.13 are based on age determinations on rocks containing radioactive materials that decay at a constant rate. By careful measurement of the components produced by radioactive decay, rocks can be dated with a great deal of precision.

Precambrian rocks, representing about 87% of earth history as we know it today, are those that were formed prior to the beginning of the Cambrian period. In table 1.13 we have broken the Precambrian into two eons, the *Archean* and the *Proterozoic*. While the majority of Precambrian rocks are igneous or metamorphic in type and the field relationships where they are exposed are often complex, undeformed sedimentary rocks of Precambrian age are known from a variety of localities.

The geologic time scale is introduced here primarily to provide a broader context in which the development of simple geologic columns representing very small segments of geologic time can be placed. Other information in table 1.13, such as the abbreviations of geologic time units and standard colors used on geologic maps, is for general information only, and has no application here. However, in Part 4 where geologic maps are considered in detail, the use of these abbreviations and map colors will become apparent.

TABLE 1.13 *Geologic Time Scale as Used in North America. (Absolute ages based on Geological Society of America, 1999 Geologic Time Scale.)*

EON	ERA	PERIOD	EPOCH	APPROXIMATE AGE IN MILLIONS OF YEARS BEFORE PRESENT	MAP SYMBOL	COMMON MAP COLOR
Phanerozoic	CENOZOIC	*Quaternary*	Recent (Holocene)	0.01	Q	Various shades of gray and yellow
			Pleistocene	1.8	Q	
		Tertiary	Pliocene	5.3	Pl or Tpl	Various shades of orange, yellow-orange, and yellow
			Miocene	23.8	M or Tm	
			Oligocene	33.7	ɸ or To	
			Eocene	54.8	E or Te	
			Paleocene	65	Tp	
	MESOZOIC	*Cretaceous*		144	K	Various shades of green
		Jurassic		206	J	Various shades of blue-green
		Triassic		248	T̄R	Various shades of blue
	PALEOZOIC	*Permian*		290	P or Cpm	Commonly blue, green, purple, pink, lavender, purple-gray
		Pennsylvanian		323	ℙ or Cp	
		Mississippian		354	M or Cm	
		Devonian		417	D	Various shades of purple, pink, lavender, tan, brown, red-brown, red
		Silurian		443	S	
		Ordovician		490	O	
		Cambrian		543	€	
Proterozoic	PRECAMBRIAN			2,500	p€	No standard color
Archean			Origin of earth	4,500?		

We will now consider the means by which a geologic column for a given locality is constructed.

Constructing a Geologic Column

Figure 1.63 shows a series of rock strata in a *geologic cross section* with a corresponding *geologic column* to the right. The geologic column is constructed by applying the law of superposition to the cross section. It is apparent that the limestone in figure 1.63 is the oldest, the shale is younger than the limestone, and the sandstone is the youngest of the three formations. The process of sedimentation was continuous during the time it took for these three layers to be deposited; only the depositional environment changed. Figure 1.64 shows the symbols used to portray various rock types on geologic cross sections and geologic columns.

Unconformities

From the information available in figure 1.63, one can conclude that some time after the sandstone was formed, the marine environment in which it formed changed to an environment of erosion or nondeposition. One cannot deduce from figure 1.63 whether additional strata were laid on top of the sandstone. All that can be said is that a depositional environment gave way to an erosional one some time after the sandstone was formed. Now, if the sea that once covered the area underlain by the three strata in figure 1.63 invaded the area again, and another sequence of strata was deposited, the situation would be depicted as in figure 1.65. The old erosional surface that is now buried beneath the younger strata is called an *unconformity* or a *surface of erosion.* It constitutes an unknown amount of geologic time that elapsed between the cessation of deposition of the older sequence of limestone-shale-sandstone and the upper sequence of conglomerate-arkose-siltstone.

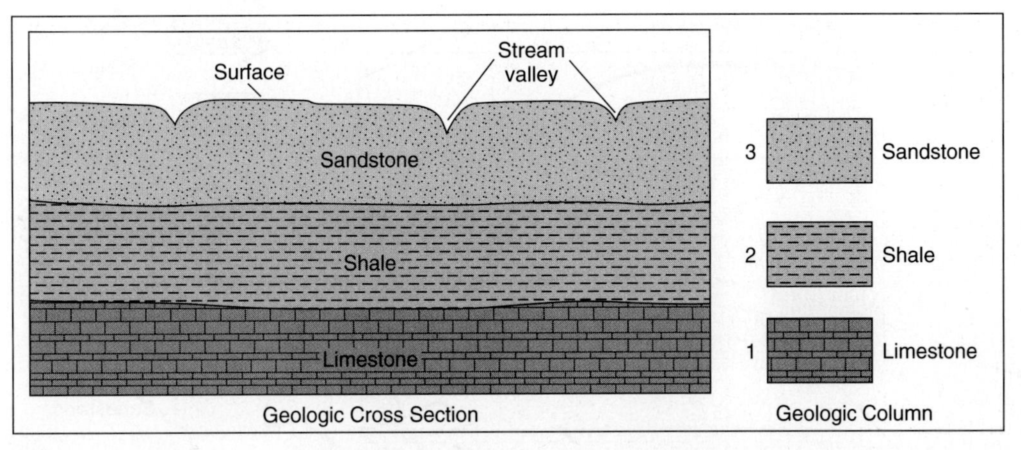

FIGURE 1.63

A simple geologic cross section and corresponding geologic column showing the relative ages of the three strata with the oldest (1) at the bottom and the youngest (3) at the top. Construction of the geologic column is based on the law of superposition.

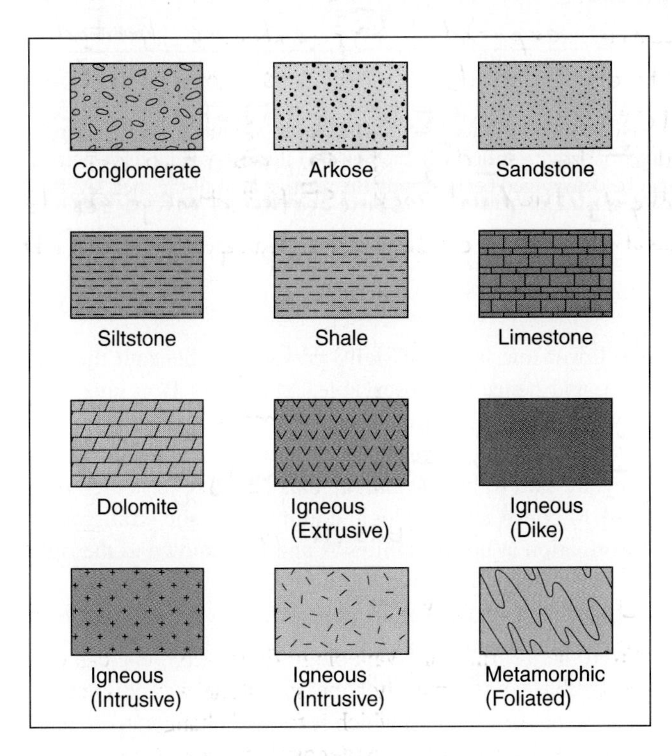

FIGURE 1.64

Symbols used in geologic columns and geologic cross sections. The colors are used to differentiate rock types only and bear no relationship to the colors used on geologic maps listed in table 1.13.

Three types of unconformities are recognized by geologists. The type shown in figures 1.65 and 1.66 is a *disconformity,* a surface that represents missing rock strata but the beds above and below are parallel to one another. A second type is a *nonconformity* (fig. 1.67), an unconformity in which an erosion surface on plutonic or metamorphic rocks has been covered by younger sedimentary or volcanic beds. A third type is an *angular unconformity* (fig. 1.68). This type of unconformity records a period of erosion of folded or tilted rocks followed by deposition of younger flat-lying sedimentary rocks.

In figure 1.65, the law of superposition still applies in constructing the geologic column. The relative ages of all six formations and the disconformity are shown by the appropriate numbers next to the boxes in the geologic column. The disconformity represents a break of unknown geologic duration in the sedimentary record for this particular locality. It is assumed that this disconformity represents a period of erosion, but the actual length of time in years represented by the unconformity cannot be determined from the information in figure 1.65. All that can be done is to place it in its relative chronological position in the geologic column.

CORRELATION

In figure 1.65, the two periods of sedimentation and the disconformity between them all represent the passage of

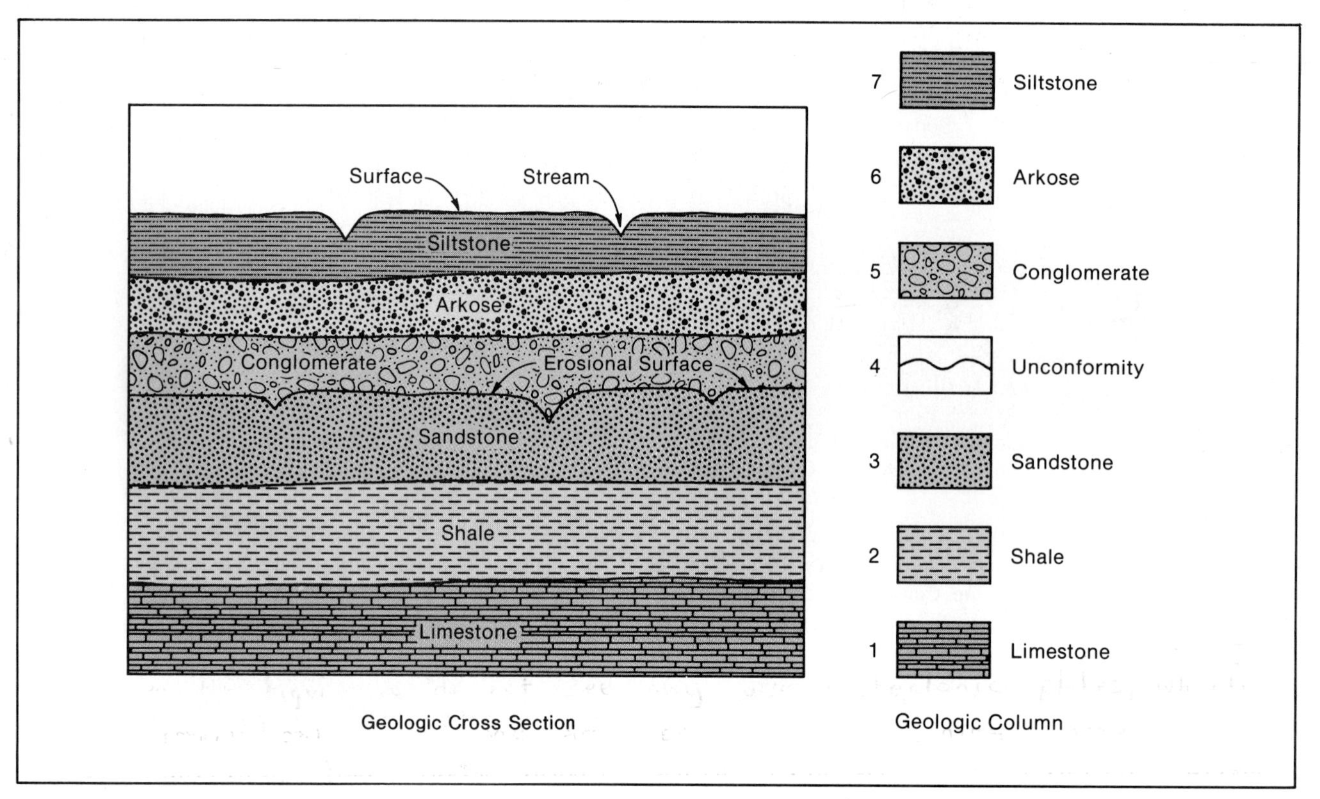

FIGURE 1.65

Geologic cross section showing two sequences of sedimentary strata separated by an unconformity. The geologic column at the right shows the sedimentary layers and unconformity of the cross section arranged in chronological order with the oldest at the bottom and the youngest at the top. The geologic time encompassed by the geologic column cannot be determined because only the relative ages of the rock layers can be deduced from the geologic cross section.

geologic time in the same geographic locality. A geologic column constructed therefrom is based only on the rocks that crop out in that particular area. However, by tracing certain formations from this locality to others nearby, it may be possible to extend the geologic column to include rocks older or younger.

As an example, consider figure 1.66 in which two geologic cross sections are shown from two locations separated by a few miles. Locality A contains a sedimentary sequence with six lithologically distinct formations. At locality B, only the upper sedimentary sequence crops out; the lower one is out of sight, presumably beneath the surface, and hence not observable. At locality B, a lava flow lies on top of a siltstone. If, in fact, the siltstone at locality A and the siltstone at B are one and the same formation, they are said to be *correlated.* This being so, it is then possible to construct a geologic column from the stratigraphic information at both localities A and B as shown to the right in figure 1.66. The lava is the youngest formation in the column, and two unconformities are present, one between the siltstone and the lava flow, and one between the conglomerate and the sandstone.

WEB CONNECTIONS

GEOLOGIC TIME

http://pubs.usgs.gov/gip/geotime/
http://pubs.usgs.gov/gip/fossils/

The online edition of the USGS general interest publication discussing geologic time. Includes discussion of geologic time, a relative time scale, a radiometric time scale, index fossils, and the age of the earth.

PART TWO

TOPOGRAPHIC MAPS, AERIAL PHOTOGRAPHS, AND OTHER IMAGERY FROM REMOTE SENSING

BACKGROUND

The earth is the laboratory of geologists. They are interested not only in the materials of which the earth is made but also in the configuration of its surface. Two important tools are used by geologists: maps and ordinary photographs supplemented by other remote sensing imagery.

A *map* is a representation of part of the earth's surface. Maps that show only the horizontal distribution of earth features and the human-made structures are of limited use to the geologist because geologic phenomena are three-dimensional. Therefore, a map that portrays the earth's surface in three dimensions is of particular value to geologists. Maps that meet this requirement are topographic maps. The first section in Part 2 of this manual deals with topographic maps.

Remote sensing is a process whereby the image of a feature is recorded by a camera or other device and reproduced in one form or another as a "picture" of the feature. One of the oldest tools of remote sensing is the camera, which produces images on a photosensitive film. When developed by chemical means, the so-called black-and-white photographs or true color photographs are the results.

The space age saw the introduction of many other kinds of remote sensing devices that produce new kinds of images such as "radar photographs," false color images, and a variety of other "pictures." These are useful not only to geologists but also to geographers, ecologists, foresters, soil scientists, meteorologists, and the like. The radar picture or image, for example, is made by recording the radiation from earth features in a way that can be resolved into a photolike picture. Radar images from earth-orbiting satellites are used extensively to show cloud cover on televised weather reports and forecasts over most commercial television stations. Radar images made from aircraft are useful in the study of many geologic phenomena.

False color photos or images are made with a device that records infrared radiation from the earth. The resulting image shows earth features in colors that are different from the true colors. For example, vegetation shows up as red instead of green on the false color images. False color pictures generally enhance the differences in earth features due to variations in vegetation, soil, water, and rock types.

Photographs or other types of images made by cameras or other sensing devices installed in airplanes, manned spacecraft, or satellites are thus another kind of map that provide geologists with useful tools for analyzing and interpreting the components of a given landscape on earth as well as on the moon and on other planets in the solar system.

The second section of Part 2 introduces students to aerial photographs and false color images. (A radar map will be introduced in Part 4.) For those interested in further pursuing the subject of remote sensing, attention is directed to the list of references at the end of Part 2.

The goal of Part 2 of this manual is to provide students with a rudimentary knowledge of topographic maps, aerial photographs, and false color images so that they will be able to apply this knowledge in their study of landforms presented in Part 3.

MAP COORDINATES AND LAND DIVISIONS

The earth's surface is arbitrarily divided into a system of reference coordinates called *latitude* and *longitude*. This coordinate system consists of imaginary lines on the earth's surface called *parallels* and *meridians* (fig. 2.1). Both of these are best described by assuming the earth to be represented by a globe with an axis of rotation passing through the North and South poles. Meridians are circles drawn on this globe that pass through the two poles. Meridians are labeled according to their positions, in degrees, from the zero meridian, which by international agreement passes through Greenwich near London, England. The zero meridian is commonly referred to as the *Greenwich meridian*. If meridional lines are drawn for each degree in an easterly direction from Greenwich (toward Asia) and in a westerly direction from Greenwich (toward North America), a family of great circles will be created. Each one of the great circle lines of *longitude* is labeled according to the number of degrees it lies east or west of the Greenwich or zero meridian. The 180º west meridian and the 180º east meridian are one and the same great circle, and constitute the International Date Line.

A great circle represents the intersection of a plane that connects two points on the surface of a sphere and passes through the center of the sphere, in this case, the earth. The intersection of the plane with the surface divides the earth into two equal halves—hemispheres—and the arc of the great circle is the shortest distance between two points on the spherical earth.

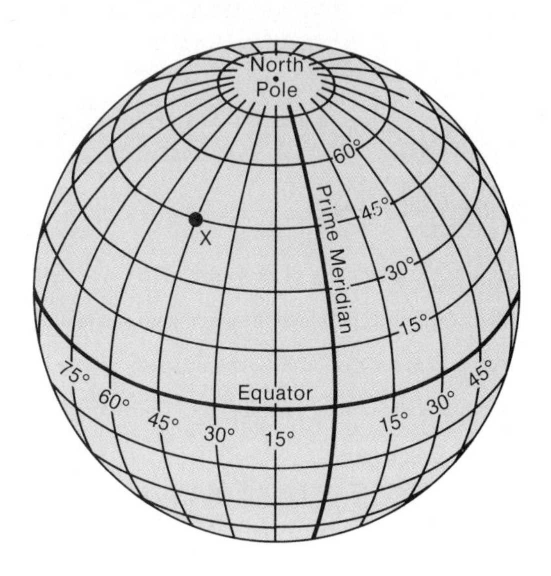

FIGURE 2.1

Generalized system of meridians and parallels. The location of point X is longitude 45° West, latitude 45° North.

Another great circle passing around the earth midway between the two poles is the equator. It divides the earth into the Northern and Southern Hemispheres. A family of lines drawn on the globe parallel to the equator constitutes the second set of reference lines needed to locate a point on the earth accurately. These lines form circles that are called *parallels of latitude*, labeled according to their distances in degrees north or south of the equator. The parallel that lies halfway between the equator and the North Pole is latitude 45° North, and the North Pole itself lies at latitude 90° North.

This system of meridians and parallels thus provides a means of accurately designating the location of any point on the globe. Santa Monica, California, for example, lies at about longitude 118° 29′ West and latitude 34° 01′ North. For increased accuracy in locating a point, degrees may be subdivided into 60 subdivisions known as *minutes*, indicated by the notation ′. Minutes may be subdivided into 60 subdivisions known as *seconds,* indicated by the notation ″. Thus, a position description might read 64° 32′ 32″ East, 44° 16′ 18″ South.

Meridional lines converge toward the North or South Pole from the equator, and the length of a degree of longitude varies from 69.17 statute miles at the equator to zero at the poles. Latitudinal lines, on the other hand, are always parallel to each other. However, because the earth is not a perfect sphere but is slightly bulged at the equator, a degree of latitude varies from 68.7 statute miles at the equator to 69.4 statute miles at the poles. Thus, the area bounded by parallels and meridians is not a true rectangle. United States Geological Survey (U.S.G.S.) quadrangle maps are also bounded by meridians and parallels, but on the scale at which they are drawn, the convergence of the meridional lines is so slight that the maps appear to be true rectangles. The U.S.G.S. standard quadrangle maps embrace an area bounded by 7½ minutes of longitude and 7½ minutes of latitude. These quadrangle maps are called 7½-minute quadrangles. Other maps published by the U.S.G.S. are 15-minute quadrangles, and a few of the older ones are 30-minute quadrangles.

Meridians always lie in a true north-south direction, and parallels always lie in a true east-west direction. *Magnetic north*, however, is the direction toward which the north-seeking end of a magnetic compass needle will point. Because the magnetic poles are not coincident with the north and south ends of the earth's rotational axis, magnetic north is different from true north except on the meridian that passes through the magnetic North Pole. The angle between true north and magnetic north is called the *magnetic declination* and is normally shown on the lower margin of most U.S.G.S. maps for the benefit of those who use a compass in the field to plot geological or other data on a base map (e.g., a U.S.G.S. standard quadrangle map).

Map Projection

While it is easy to draw the system of meridians and parallels on a globe, it is not possible to draw this system on a flat piece of paper (a map) without introducing some distortion. A wide variety of methods have been developed to reduce this distortion during the construction of a map. This process of constructing a map, the transferring of the meridians and parallels to a flat sheet of paper, is a geometric exercise called *projection,* and the resulting product is called a *map projection*. The map projection selected for use by the cartographer, one who draws maps, depends on the purpose of the map and the material to be presented. (See Box 2.1.) Most of the U.S.G.S. maps found in this manual are drawn on a polyconic projection. This projection preserves neither shape nor area, but for the small area represented the distortion of both is at a minimum. The 1:250,000 scale map of Greenwood (fig. 3.14) uses a transverse Mercator projection, and figure 4.36 uses a Lambert conformal conic projection.The map used for figure 5.1 is a Mercator projection.

Web Connections

http://www.utexas.edu/depts/grg/gcraft/notes/mapproj/mapproj_ftoc.html

A map projection overview. Very useful to view the variety of map projections available, their uses, and their limitations.

http://everest.hunter.cuny.edu/mp/

Map Projection Home Page. A collection of information relating to map projections.

Box 2.1
Map Projections

A map should show the spatial relationships of features at Earth's surface as accurately as possible in terms of distance, area, and direction. These requirements can be met when the cartographer is dealing with a relatively small area, such as a county or a small state in the United States. Over this area, Earth's curvature is minimal and the surfaces can be mapped onto a piece of flat paper with little distortion. However, at a global scale, the problems are acute; the skin of a sphere will simply not lie flat without being distorted in some way. No flat map can portray shape, distance, area, and direction over the spherical globe accurately. The different portraits of Earth shown in Box Figure 2 are different map projections, each of which attempts to minimize overall distortion or to maximize the accuracy of one measure of space. There are no right or wrong map projections, because they all contain distortion of one sort or another. In virtually all projections, linear scale is particularly not constant across the map, meaning that measurement of long distances will be inaccurate.

The simplest way to imagine constructing a map projection is to think of a transparent glove with a lightbulb inside it and a sheet of paper touching the globe in one of the three ways shown in Box Figure 1. The outline of the continents and the lines of latitude and longitude are silhouetted on the paper to form the map. In fact, most projections are not constructed as simply as this. They are calculated mathematically and may be a compromise among a number of methods and designed to preserve the best features of each. There are three main desirable properties in a map and different projections attempt to preserve each one.

Equal-area projections, as the name suggests, maintain the areas of the continents in their correct proportions across the globe. They are used to depict distributions such as population, zones of climate, soils, or vegetation. The Robinson projection used for world maps is an equal-area projection. The scale distortion in *conformal maps* is equal in the two main directions from the projection's origin but increases away from the origin. This means that conformal projections maintain the correct shape over small areas, but the outline of continents and oceans mapped over a larger area is distorted. *Azimuthal maps* are constructed around a point or *focus.* Lines of constant bearing or

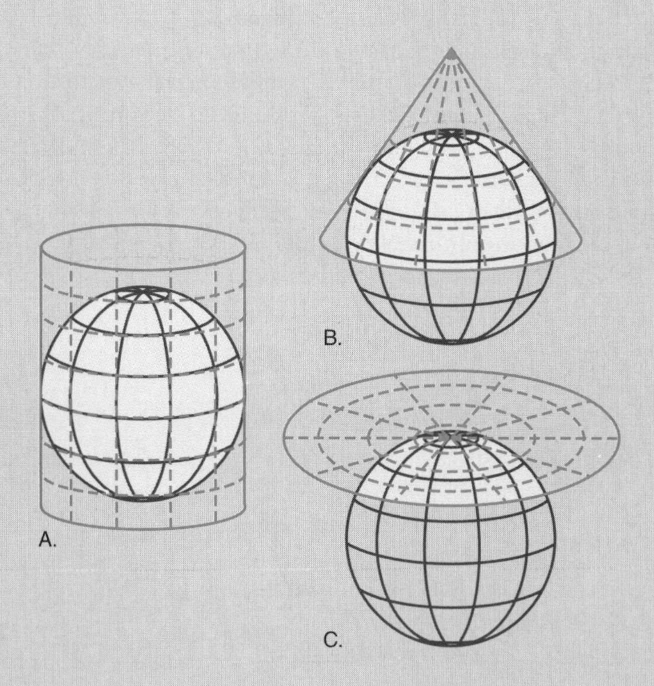

Box Figure 1

Methods of constructing map projections. (*A*) Cylindrical. (*B*) Conical. (*C*) Planar. The cylindrical and conical projections are "unwrapped" from the globe to give a flat map.

compass direction radiating from the focus are straight lines on an azimuthal map. Distortion of shape and area is symmetrical around the central point and increases away from it. The azimuthal projection is used most often in geography to portray the polar regions, which are distorted in projections designed for lower latitudes. Azimuthal maps can also be equal-area or conformal.

It is mathematically impossible to combine the properties of equal area and reasonably correct shape on one map. The commonly used Mercator projection is a conformal map and demonstrates graphically how poor this projection is for showing geographical distributions. The Mercator projection was developed in 1569 by a Flemish cartographer, Gerardus Mercator, to help explorers and navigators. It has the important property, for navigation, that a straight line drawn anywhere on the map, in any direction, gives the true compass bearing between these two points. However, the size of land masses in the mid- and high-latitudes (toward the poles) is grossly distorted in this projection. Alaska, for example, appears to be the same size as Brazil, although Brazil is actually five times as large. Greenland appears similar in size to the whole of South America. This distortion is partly because the meridians, which actually come together toward the poles, are shown with uniform spacing throughout the Mercator map. Despite its unsuitability for most geographical purposes, the Mercator is still very widely used.

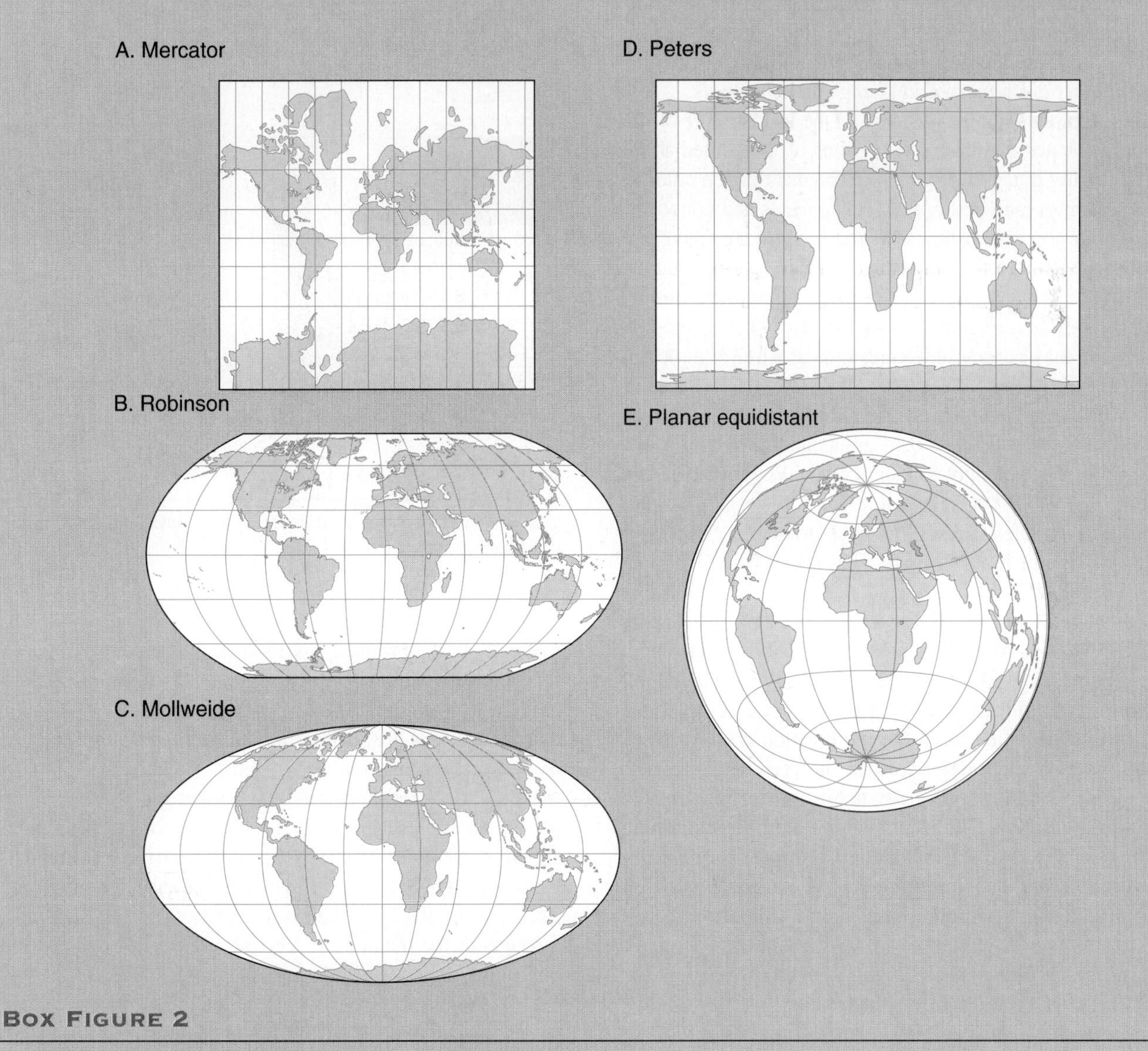

Box Figure 2

From Michael Bradshaw and Ruth Weaver, Foundations of Physical Geography.

RANGE AND TOWNSHIP

For purposes of locating property lines and land descriptions on legal documents, another system of coordinates is used in the United States and some parts of Canada. (See inside front cover.) In the U.S. this grid system is called the U.S. Public Land Survey (U.S.P.L.S.). This system is tied into the latitude and longitude coordinate system but functions independently of it. The basic block of this system is the *section,* a rectangular block of land one mile long and one mile wide. An area containing 36 sections is called a *township.* Each township consists of 36 one-square-mile sections, 6 sections on a side that are numbered according to the system shown in figure 2.2. One section of land contains 640 acres. While it was the intent of the original government land surveyors to make each section an exact square of land, many sections and townships are irregular in shape because of surveying errors and other discrepancies in laying out the network.

Examination of the maps on the inside cover of this manual indicates that there are portions of the United States that do not use this grid system. The original thirteen colonies had a system of land description, called metes and bounds, in place before the establishment of the U.S. Public Land Survey, and that system has continued. In this type of land description the starting point and bearings along the sides are given as well as a statement of the markers used at the corners, which may be natural features such as a tree or a rock. A sketch map drawn to scale is often attached to the description. The other major exception to the U.S.P.L.S. is the State of Texas where early Spanish land grants and later land grants made by the Mexican government and the Republic of Texas to settlers in the 1880s have resulted in a description system that is quite complex and difficult to describe.

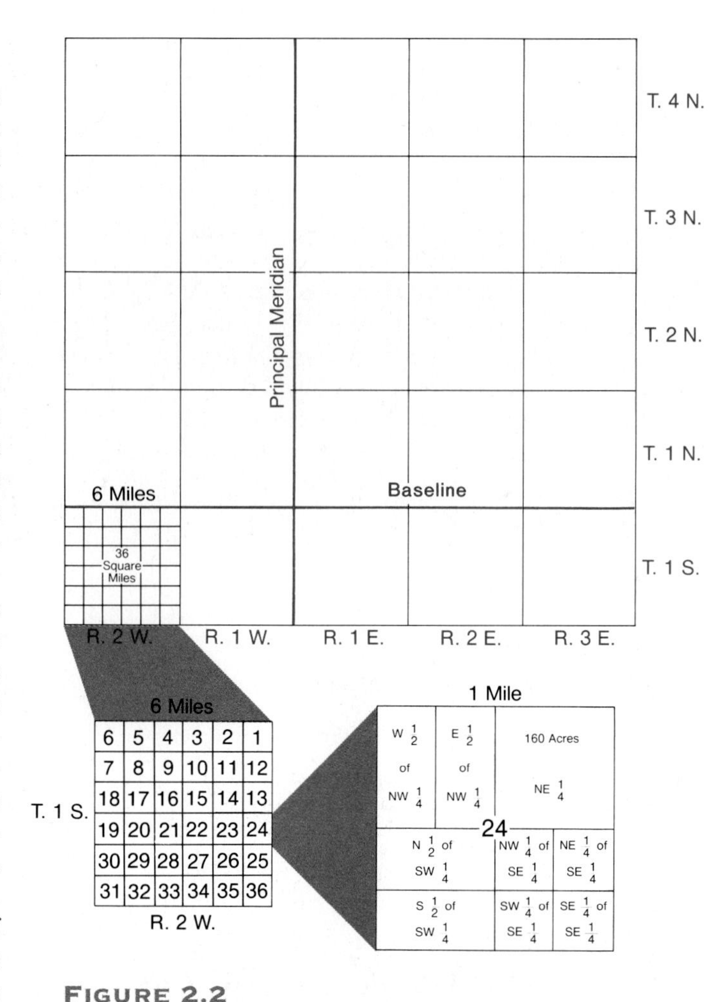

FIGURE 2.2

Standard land divisions used in the United States and some parts of Canada.

The north-south lines marking township boundaries are called *range lines,* and the east-west boundaries are called *township lines.* The coordinate system of numbering townships has a reference or beginning point the intersection of a meridian of longitude (the *principal meridian*) and a parallel of latitude (the baseline). A particular township is identified by stating its position north or south of the baseline and east or west of the principal meridian. The system of numbering township and range lines is shown in figure 2.2. The letter *T* along the right-hand margin of the large map stands for the word *township,* and the letter *R* stands for the word *range.* The notation T. 1 S, R. 2 W. is read, "Township one south, Range two west." Under this system, each township has a unique numerical designation.

In the U.S., while most states have a separate prime meridian and base line, several states share them with adjoining states so that there are only 34 U.S.P.L.S. systems. Several principal meridians and baselines are used in the coterminous United States, so that the township and range coordinate numbers are never very large.

For purposes of locating either human-made or natural features in a given section, an additional convention is employed. This consists of dividing the section into quarters called the northeast quarter (NE¼), the southwest quarter (SW¼), and so on. Sections may also be divided into halves such as the north half (N½) or west half (W½). The quarter sections are further divided into four more quarters or two halves, depending on how refined one wants to make the description of a feature on the ground. For example, an exact description of the 40 acres of land in the extreme southeast corner of the map of section 24 in figure 2.2 would be as follows: SE¼ of the SE¼ of Section 24, T. 1 S., R. 2 W. This style of notation will be used throughout the manual in referring the student to a particular feature on a map used in an exercise.

Figure 2.3 is an aerial photograph of rural Iowa. The rectangular network of roads tends to follow section lines, and the cultivated fields generally conform in shape and size according to the system of land divisions shown in figure 2.2.

Because the maps used in this manual are only **selected parts** of standard U.S.G.S. quadrangle maps, the township and range lines numbering system, magnetic declination, and other data usually found on the lower margin of the maps may not be included. Where necessary, these data will be supplied in the text of the exercise.

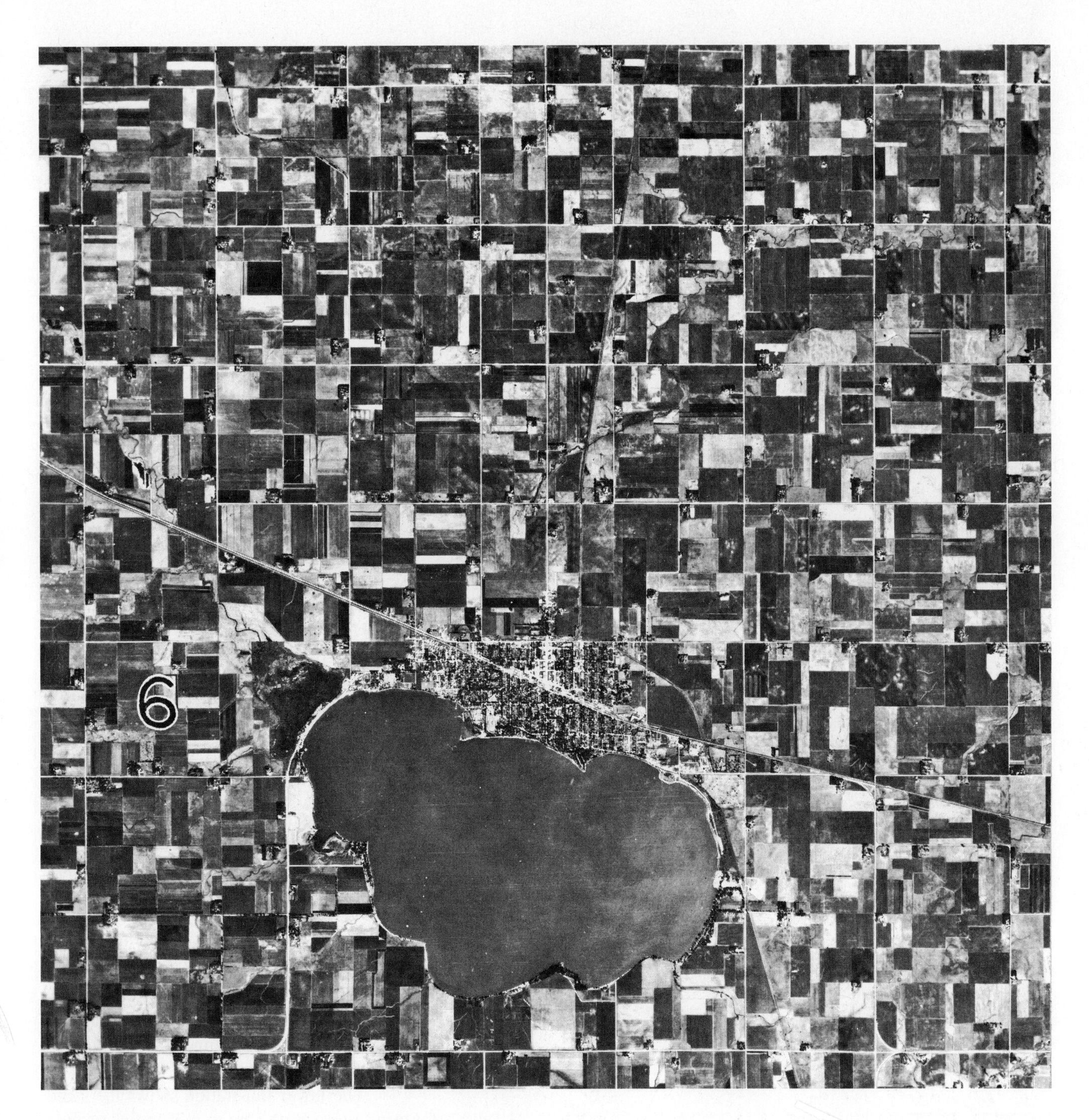

FIGURE 2.3

Aerial photograph of a rural area in Iowa. The town is Storm Lake.

(Courtesy of the U.S.G.S. Photo was taken on September 23, 1950.)

Topographic Maps

Definition

A *topographic* map is a graphic representation of the three-dimensional configuration of the earth's surface. Most topographic maps also show land boundaries and other human-made features. The United States Geological Survey (U.S.G.S.), a unit of the Department of the Interior, has been actively engaged in the making of a series of standard topographic maps of the United States and its possessions since 1882.

Features of Topographic Maps

The features shown on topographic maps may be divided into three major groups: (1) *topography* or relief (printed in brown), depicted by the configuration of contour lines that show hills, valleys, mountains, plains, and the like; (2) *water features,* (printed in blue), including oceans, lakes, ponds, rivers, canals, swamps, and the like; (3) *culture,* (printed in black), representing human-made works such as roads, railroads, prominent buildings, land boundaries, and similar features (fig. 2.4). Geographical names are also printed in black.

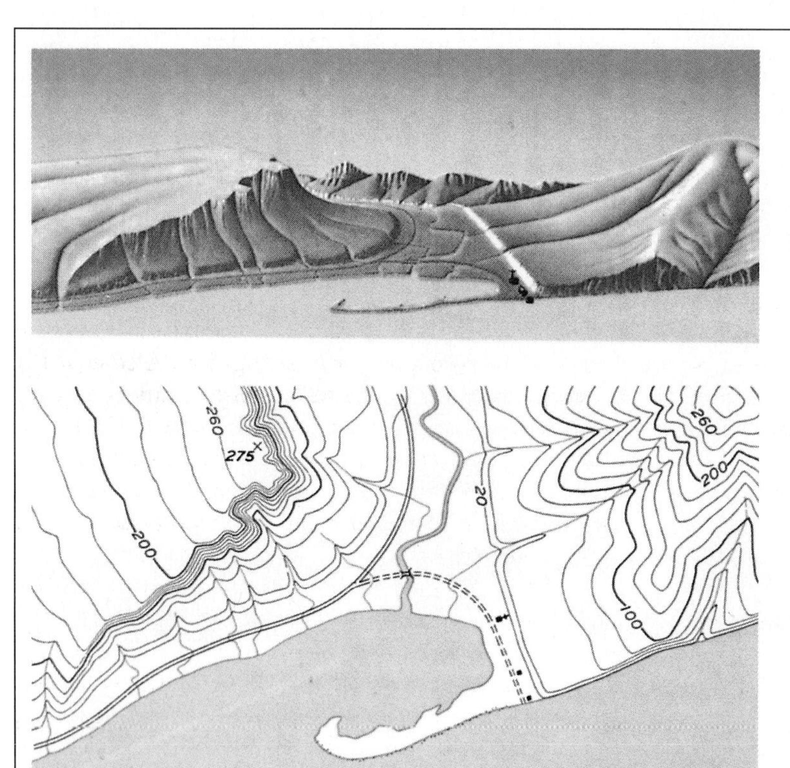

The Use of Symbols in Mapping

These illustrations show how various features are depicted on a topographic map. The upper illustration is a perspective view of a river valley and the adjoining hills. The river flows into a bay that is partly enclosed by a hooked sandbar. On either side of the valley are terraces through which streams have cut gullies. The hill on the right has a smoothly eroded form and gradual slopes, whereas the one on the left rises abruptly in a sharp precipice from which it slopes gently and forms an inclined table land traversed by a few shallow gullies. A road provides access to a church and two houses situated across the river from a highway that follows the seacoast and curves up the river valley.

The lower illustration shows the same features represented by symbols on a topographic map. The contour interval (the vertical distance between adjacent contours) is 20 feet.

Figure 2.4

Landforms as shown on a topographic map. Contour lines are printed in brown, water features in blue, and human-made structures in black.

(Courtesy of the U.S.G.S.)

On some topographic maps, woodland cover (forests, orchards, vineyards, and scrub) are shown in green (see fig. 3.31); important roads and public land surveys in red (fig. 3.14); and where existing maps have been corrected through the use of aerial photos without field checks, the added features are shown in purple.

Figure 2.5 provides detailed information about the standard symbols used on topographic maps published by the U.S.G.S. Additional information about U.S.G.S. maps is provided with figure 2.5.

Standard topographic maps of the U.S.G.S. cover a *quadrangle* of area that is bounded by *parallels of latitude* (forming the northern and southern margins of the map) and by *meridians of longitude* (forming the eastern and western margins of the map). The published maps have different *scales.* A map scale is a means of showing the relationship between the size of an object or feature indicated on a map and the corresponding actual size of the same object or feature on the ground.

Geologists make use of topographic maps because they provide them with a means to observe earth features in *three dimensions.* Unlike other maps, topographic maps show natural features to a fair degree of accuracy in terms of length, width, and vertical height or depth. Thus by examination of a topographic map, and through an understanding of the symbols shown thereon, geologists are able to interpret earth features and draw conclusions as to their origin in the light of geologic processes.

In the United States, Canada, and other English-speaking countries of the world, most maps produced to date have used the English system of measurement. That is to say, distances are measured in feet, yards, or miles; elevations are shown in feet; and water depths are recorded in feet or fathoms (1 fathom = 6 feet). In 1977, in accordance with national policy, the U.S.G.S. formally announced its intent to convert all of its maps to the metric system. As resources and circumstances permit, new maps published by the U.S.G.S. will show distances in kilometers and elevations in meters. The conversion from English to metric units will take many decades in the United States. In this manual, most maps used will be those published by the U.S.G.S. *prior* to the adoption of the metric system, because the metric maps are still insufficient in number to portray the great diversity of geologic features presented in this manual. However, some examples of map scales in metric units will be given to acquaint students with the system.

To help students familiarize themselves with the metric system and its relationship to the English system, table 2.1 is provided as a convenient reference.

TABLE 2.1 *Units of Measurement in the English and Metric Systems, and the Means of Converting from One to the Other*

A. English Units of Linear Measurement
12 inches = 1 foot
3 feet = 1 yard
1 mile = 1,760 yards; 5,280 feet; 63,360 inches

B. Metric Units of Linear Measurement
10 millimeters = 1 centimeter
100 centimeters = 1 meter
1,000 meters = 1 kilometer

C. Conversion of English Units to Metric Units

symbol	when you know	multiply by	to find	symbol
in.	inches	2.54	centimeters	cm
ft.	feet	30.48	centimeters	cm
yd.	yards	0.91	meters	m
mi.	miles	1.61	kilometers	km

D. Conversion of Metric Units to English Units

symbol	when you know	multiply by	to find	symbol
mm	millimeters	0.04	inches	in.
cm	centimeters	0.4	inches	in.
m	meters	3.28	feet	ft.
m	meters	1.09	yards	yd.
km	kilometers	0.62	miles	mi.

ELEMENTS OF A TOPOGRAPHIC MAP

TOPOGRAPHY

Topography is the configuration of the land surface and is shown by means of *contour lines* (fig. 2.4). A contour line is an imaginary line on the surface of the earth connecting points of equal elevation. The *contour interval* (C.I.) is the difference in elevation of any two adjacent contour lines. Elevations are given in feet or meters above mean sea level. The shore of a lake is, in effect, a contour line because every point on it is at the same level (elevation).

Contour lines are brown on the standard U.S.G.S. maps. The C.I. is usually constant for a given map and may range from 5 feet for flat terrain to 50 or 100 feet for a mountainous region. C.I.'s for U.S.G.S. maps using the metric system are 1, 2, 5, 10, 20, 50, or 100 meters, depending upon the smoothness or ruggedness of the terrain to be depicted. Usually, every fifth contour line is printed in a heavier line than the others and bears the elevation of the contour above sea level. In addition to contour lines, the heights of many points on the map, such as road intersections, summits

CONTROL DATA AND MONUMENTS

Aerial photograph roll and frame number* 3-20

Horizontal control

- Third order or better, permanent mark — Neace
- With third order or better elevation — BM 45.1; Pike BM 45.1
- Checked spot elevation — 19.5
- Coincident with section corner — Cactus
- Unmonumented*

Vertical control

- Third order or better, with tablet — BM × 16.3
- Third order or better, recoverable mark — × 120.0
- Bench mark at found section corner — BM 18.6
- Spot elevation — × 5.3

Boundary monument

- With tablet — BM 21.6; BM 71
- Without tablet — 171.3
- With number and elevation — 67 301.1

U.S. mineral or location monument

CONTOURS

Topographic

- Intermediate
- Index
- Supplementary
- Depression
- Cut; fill

Bathymetric

- Intermediate
- Index
- Primary
- Index Primary
- Supplementary

BOUNDARIES

- National
- State or territorial
- County or equivalent
- Civil township or equivalent
- Incorporated city or equivalent
- Park, reservation, or monument
- Small park

*Provisional Edition maps only

Provisional Edition maps were established to expedite completion of the remaining large scale topographic quadrangles of the conterminous United States. They contain essentially the same level of information as the standard series maps. This series can be easily recongnized by the title "Provisional Edition" in the lower right hand corner.

LAND SURVEY SYSTEMS

U.S. Public Land Survey System

- Township or range line
  - Location doubtful
- Section line
  - Location doubtful

Found section corner; found closing corner

Witness corner; meander corner — WC; MC

Other land surveys

- Township or range line
- Section line

Land grant or mining claim; monument

Fence line

SURFACE FEATURES

- Levee — Levee
- Sand or mud area, dunes, or shifting sand — Sand
- Intricate surface area — Strip mine
- Gravel beach or glacial moraine — Gravel
- Tailings pond — Tailings Pond

MINES AND CAVES

- Quarry or open pit mine
- Gravel, sand, clay, or borrow pit
- Mine tunnel or cave entrance
- Prospect; mine shaft — X
- Mine dump — Mine dump
- Tailings — Tailings

VEGETATION

- Woods
- Scrub
- Orchard
- Vineyard
- Mangrove — Mangrove

GLACIERS AND PERMANENT SNOWFIELDS

- Contours and limits
- Form lines

MARINE SHORELINE

Topographic maps

- Approximate mean high water
- Indefinite or unsurveyed

Topographic-bathymetric maps

- Mean high water
- Apparent (edge of vegetation)

FIGURE 2.5

Standard symbols used on topographic maps published by the U.S.G.S.

(Courtesy of the U.S.G.S.)

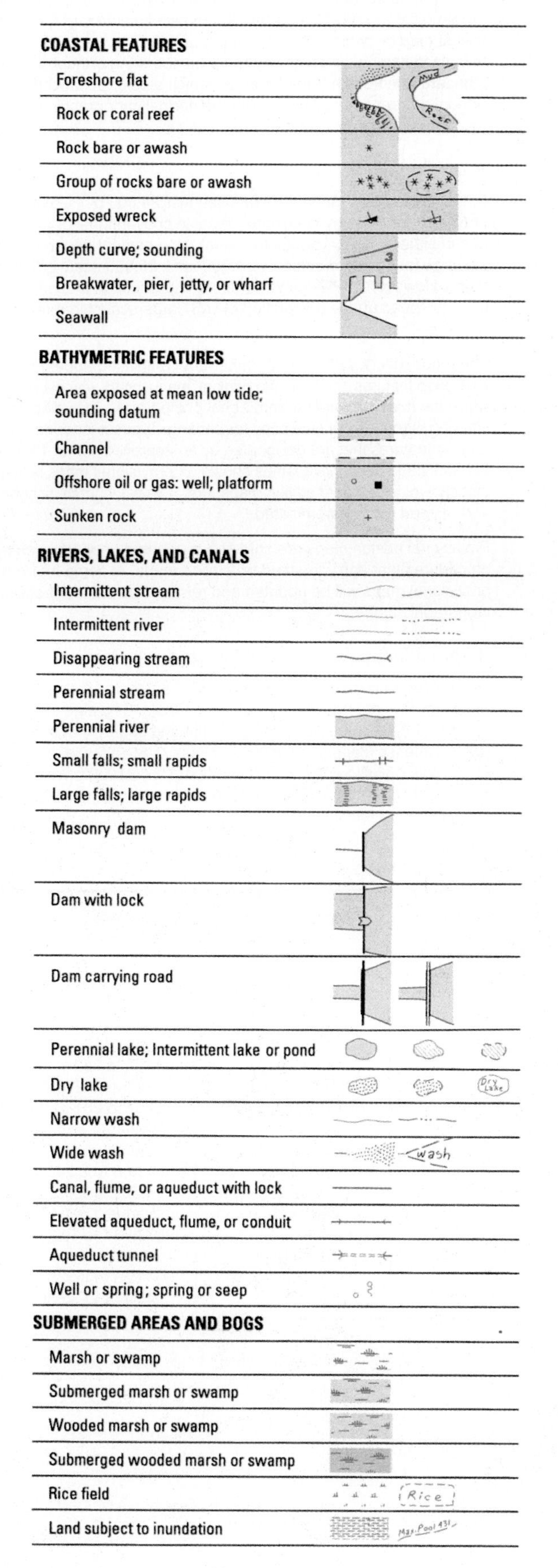

COASTAL FEATURES

- Foreshore flat
- Rock or coral reef
- Rock bare or awash
- Group of rocks bare or awash
- Exposed wreck
- Depth curve; sounding
- Breakwater, pier, jetty, or wharf
- Seawall

BATHYMETRIC FEATURES

- Area exposed at mean low tide; sounding datum
- Channel
- Offshore oil or gas: well; platform
- Sunken rock

RIVERS, LAKES, AND CANALS

- Intermittent stream
- Intermittent river
- Disappearing stream
- Perennial stream
- Perennial river
- Small falls; small rapids
- Large falls; large rapids
- Masonry dam
- Dam with lock
- Dam carrying road
- Perennial lake; Intermittent lake or pond
- Dry lake
- Narrow wash
- Wide wash
- Canal, flume, or aqueduct with lock
- Elevated aqueduct, flume, or conduit
- Aqueduct tunnel
- Well or spring; spring or seep

SUBMERGED AREAS AND BOGS

- Marsh or swamp
- Submerged marsh or swamp
- Wooded marsh or swamp
- Submerged wooded marsh or swamp
- Rice field
- Land subject to inundation

BUILDINGS AND RELATED FEATURES

- Building
- School; church
- Built-up Area
- Racetrack
- Airport
- Landing strip
- Well (other than water); windmill
- Tanks
- Covered reservoir
- Gaging station
- Landmark object (feature as labeled)
- Campground; picnic area
- Cemetery: small; large — Cem

ROADS AND RELATED FEATURES

Roads on Provisional edition maps are not classified as primary, secondary, or light duty. They are all symbolized as light duty roads.

- Primary highway
- Secondary highway
- Light duty road
- Unimproved road
- Trail
- Dual highway
- Dual highway with median strip
- Road under construction — U. C.
- Underpass; overpass
- Bridge
- Drawbridge
- Tunnel

RAILROADS AND RELATED FEATURES

- Standard gauge single track; station
- Standard gauge multiple track
- Abandoned
- Under construction
- Narrow gauge single track
- Narrow gauge multiple track
- Railroad in street
- Juxtaposition
- Roundhouse and turntable

TRANSMISSION LINES AND PIPELINES

- Power transmission line: pole; tower
- Telephone line — Telephone
- Aboveground oil or gas pipeline
- Underground oil or gas pipeline — Pipeline

Map series and quadrangles

Each map in a U.S. Geological Survey series conforms to established specifications for size, scale, content, and symbolization. Except for maps which are formatted on a county or state basis, U.S.G.S. quadrangle series maps cover areas bounded by parallels of latitude and meridians of longitude.

Map scale

Map scale is the relationship between distance on a map and the corresponding distance on the ground. Scale is expressed as a ratio, such as 1:25,000, and shown graphically by bar scales marked in feet and miles or in meters and kilometers.

Standard edition maps

Standard edition topographic maps are produced at 1:20,000 scale (Puerto Rico) and 1:24,000 or 1:25,000 scale (conterminous United States and Hawaii) in either 7.5 x 7.5- or 7.5 x 15-minute format. In Alaska, standard edition maps are available at 1:63,360 scale in 7.5 x 20 to 36-minute quadrangles. Generally, distances and elevations on 1:24,000-scale maps are given in conventional units: miles and feet, and on 1:25,000-scale maps in metric units: kilometers and meters.

The shape of the Earth's surface, portrayed by contours, is the distinctive characteristic of topographic maps. Contours are imaginary lines that follow the land surface or the ocean bottom at a constant elevation above or below sea level. The contour interval is the elevation difference between adjacent contour lines. The contour interval is chosen on the basis of the map scale and on the local relief. A small contour interval is used for flat areas; larger intervals are used for mountainous terrain. In very flat areas, the contour interval may not show sufficient surface detail and supplementary contours at less than the regular interval are used.

The use of color helps to distinguish kinds of features:

Black—cultural features such as roads and buildings.
Blue—hydrographic features such as lakes and rivers.
Brown—hypsographic features shown by contour lines.
Green—woodland cover, scrub, orchards, and vineyards.
Red—important roads and public land survey system.
Purple—features added from aerial photographs during map revision. The changes are not field checked.

Some quadrangles are mapped by a combination of orthophotographic images and map symbols. Orthophotographs are derived from aerial photographs by removing image displacements due to camera tilt and terrain relief variations. An orthophotoquad is a standard quadrangle format map on which an orthophotograph is combined with a grid, a few place names, and highway route numbers. An orthophotomap is a standard quadrangle format map on which a color enhanced orthophotograph is combined with the normal cartographic detail of a standard edition topographic map.

Provisional edition maps

Provisional edition maps are produced at 1:24,000 or 1:25,000 scale (1:63,360 for Alaskan 15-minute maps) in conventional or metric units and in either a 7.5 x 7.5-minute format. Map content generally is the same as for standard edition 1:24,000- or 1:25,000-scale quadrangle maps. However, modified symbolism and production procedures are used to speed up the completion of U.S. large-scale topographic map coverage.

The maps reflect a provisional rather than a finished appearance. For most map features and type, the original manuscripts which are prepared when the map is compiled from aerial photographs, including hand lettering, serve as the final copy for printing. Typeset lettering is applied only for features that are designated by an approved name. The number of names and descriptive labels shown on provisional maps is less than that shown on standard edition maps. For example, church, school, road, and railroad names are omitted.

Provisional edition maps are sold and distributed under the same procedures that apply to standard edition maps. At some future time, provisional maps will be updated and reissued as standard edition topographic maps.

National Mapping Program indexes

Indexes for each state, Puerto Rico, the U.S. Virgin Islands, Guam, American Samoa, and Antarctica are available. Separate indexes are available for 1:100,000-scale quadrangle and county maps; USGS/Defense Mapping Agency 15-minute (1:50,000-scale) maps; U.S. small-scale maps (1:250,000, 1:1,000,000, 1:2,000,000 scale; state base maps; and U.S. maps); land use/land cover products; and digital cartographic products.

Series	Scale	1 inch represents approximately	1 centimeter represents	Size (latitude x longitude)	Area (square miles)
Puerto Rico 7.5-minute	1:20,000	1,667 feet	200 meters	7.5 x 7.5 min.	71
7.5-minute	1:24,000	2,000 feet (exact)	240 meters	7.5 x 7.5 min.	49 to 70
7.5-minute	1:25,000	2,083 feet	250 meters	7.5 x 7.5 min.	49 to 70
7.5 x 15-minute	1:25,000	2,083 feet	250 meters	7.5 x 15 min.	98 to 140
USGS/DMA 15-minute	1:50,000	4,166 feet	500 meters	15 x 15 min.	197 to 282
15-minute	1:62,500	1 mile	625 meters	15 x 15 min.	197 to 282
Alaska 1:63,360	1:63,360	1 mile (exact)	633.6 meters	15 x 20 to 36 min.	207 to 281
County 1:50,000	1:50,000	4,166 feet	500 meters	County area	Varies
County 1:100,000	1:100,000	1.6 miles	1 kilometer	County area	Varies
30 x 60-minute	1:100,000	1.6 miles	1 kilometer	30 x 60 min.	1,568 to 2,240
U.S. 1:250,000	1:250,000	4 miles	2.5 kilometers	1° x 2° or 3°	4,580 to 8,669
State maps	1:500,000	8 miles	5 kilometers	State area	Varies
U.S. 1:100,000	1:1,000,000	16 miles	10 kilometers	4° x 6°	73,734 to 102,759
U.S. Sectional	1:2,000,000	32 miles	20 kilometers	State groups	Varies
Antarctica 1:250,000	1:250,000	4 miles	2.5 kilometers	1° x 3° to 15°	4,089 to 8,336
Antarctica 1:500,000	1:500,000	8 miles	5 kilometers	2° x 7.5°	28,174 to 30,462

How to order maps

Mail orders. Order by map name, state, and series/scale. Payment by money order or check payable to the U.S. Geological Survey must accompany your order. Your complete address, including ZIP code, is required.

Maps of areas *east* of the Mississippi River, including Minnesota, Puerto Rico, the Virgin Islands of the United States, and Antarctica.	Eastern Distribution Branch U.S. Geological Survey 1200 South Eads Street Arlington, VA 22202
Maps of areas *west* of the Mississippi River, including Alaska, Hawaii, Louisiana, American Samoa, and Guam.	Western Distribution Branch U.S. Geological Survey Box 25286, Federal Center Denver, CO 80225

travelers to visit. It contains many of the geological wonders of world with links to most of them. Other interesting natural phenomena are also presented.

http://info.er.usgs.gov/network/index.html

A compilation of USGS Internet resources. Provides a listing of scientific research resources, USGS Online Databases, and some World Wide Web Search Engines.

http://usgs.gov/education/

This site represents a portion of the USGS web dedicated to K-12 education, exploration, and life-long learning. Students will find some useful information here and those in science education will find it a good resource for the future.

Topographic Profiles

A *topographic profile* is a diagram that shows the change in elevation of the land surface along any given line. It represents graphically the "skyline" as viewed from a distance. Features shown in profile are viewed along a horizontal line of sight, whereas features shown on a map or in *plan view* are viewed along a vertical line of sight. Topographic profiles can be constructed from a topographic map along any given line.

The *vertical scale* of a profile is arbitrarily selected and is usually, but not always, larger than the horizontal scale of the map from which the profile is drawn. Only when the horizontal and vertical scales are the same is the profile *true,* but in order to facilitate the drawing of the profile and to emphasize differences in relief, a larger vertical scale is used. Such profiles are *exaggerated profiles.*

Ideally both the horizontal and vertical scales would be the same. This is impractical in most cases. On a section with a horizontal scale of 1:1,000,000, the topographic features would be almost impossible to see if the same vertical scale were used. Both horizontal and vertical scales are usually provided with each profile. The exaggeration is determined by comparing the inches on the profile with feet in nature. Thus, on a profile with a horizontal scale of 1:24,000 and a vertical scale of 1/10 inch to every 100 feet of elevation:

Horizontally 1 inch represents 24,000 divided by 12 = 2,000 ft.

Vertically 1 inch represents 10 times 100 = 1,000 ft.

Vertical exaggeration is 2.0 times.

Be aware that vertical exaggeration not only increases but also changes the character of the profile. A volcano such as that shown in figure 2.14 would appear as a sharp peak at 10 times vertical exaggeration.

Instructions for Drawing a Topographic Profile

Figure 2.10 shows the relationship of a topographic profile to a topographic map. It should be examined in connection with the following instructions:

1. The line along which a cross section is to be constructed may be defined by an actual line drawn on the map or by two points on the map that determine the terminals of the line of cross section.
2. Examine the line along which the profile is to be drawn and note the difference between the highest and lowest contours crossed by it. The difference between them is the maximum relief of the profile. Cross-sectional paper divided into 0.1 inch or 2 mm squares makes a good base on which to draw a profile. Use a vertical scale as small as possible so as to keep the amount of vertical exaggeration to a minimum. For example, for a profile along which the maximum relief is less than 100 feet, a vertical scale of 0.1 inch or 2 mm = 5, 10, 20, or 25 feet is appropriate. For a profile with 100 to 500 feet of maximum relief, a vertical scale of 0.1 inch or 2 mm = 40 or 50 feet is proper. If the maximum relief along the profile is between 500 and 1,000 feet, a vertical scale of 0.1 inch or 2 mm = 80 or 100 feet is adequate. When the maximum relief is greater than 1,000 feet, a vertical scale of 0.1 inch or 2 mm = 200 feet is appropriate. The general rule for guidance in the selection of a vertical scale is: **The greater the maximum relief, the smaller the scale.** Label the horizontal lines of the profile grid with appropriate elevations from the contours crossed by the line of the profile. Every other line on a 0.1 inch or 2 mm grid is sufficient.
3. Place the edge of the cross-sectional paper along the line of profile. Opposite each intersection of a contour line with the line of profile, mark a short dash at the edge of the cross-sectional paper. If the contour lines are closely spaced, only the heavy or *index* contours need to be marked. Also mark the positions of streams, lakes, hilltops, and significant cultural features on the line of profile. At the edge of the paper, label the elevation of each dash.
4. Drop these elevations perpendicularly to the corresponding elevations represented by the horizontal lines on the cross-sectional paper.
5. Connect these points by a smooth line and label significant features such as streams and summits of hills. Add the horizontal scale and write a title on the profile.

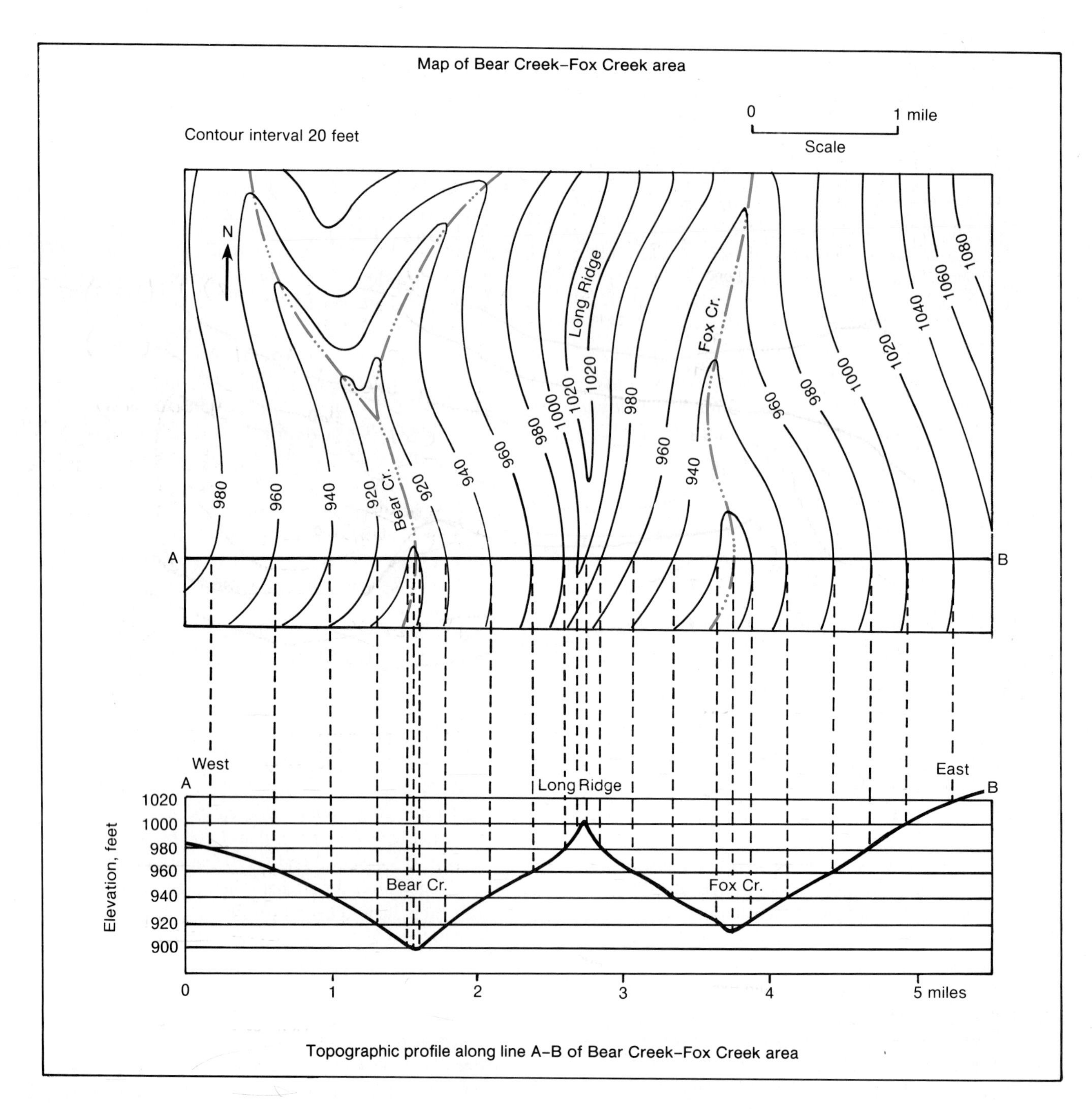

FIGURE 2.10

A topographic profile drawn along line A-B on the map of the hypothetical Bear Creek-Fox Creek area. See text for step-by-step instructions.

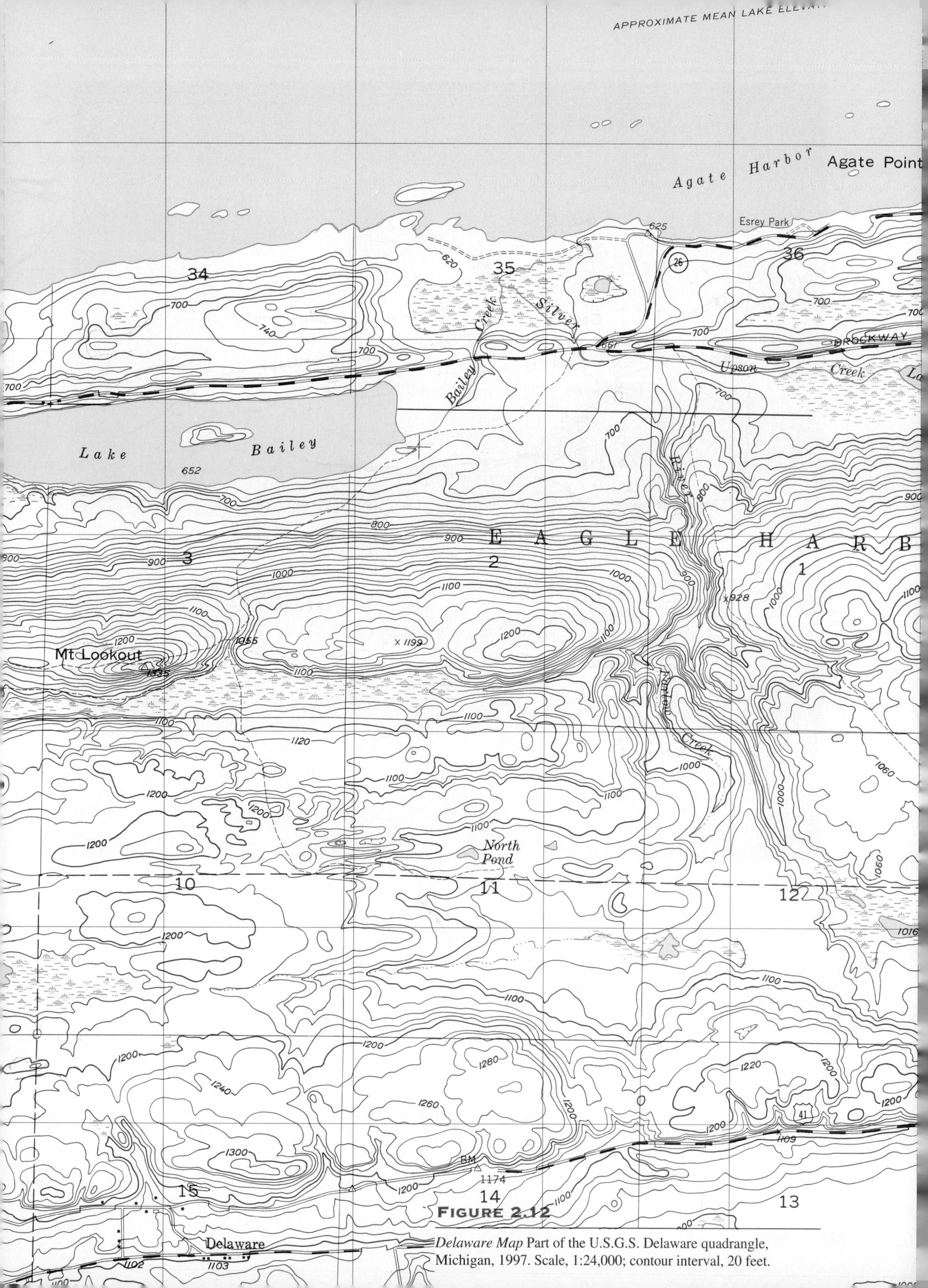

FIGURE 2.12

Delaware Map Part of the U.S.G.S. Delaware quadrangle, Michigan, 1997. Scale, 1:24,000; contour interval, 20 feet.

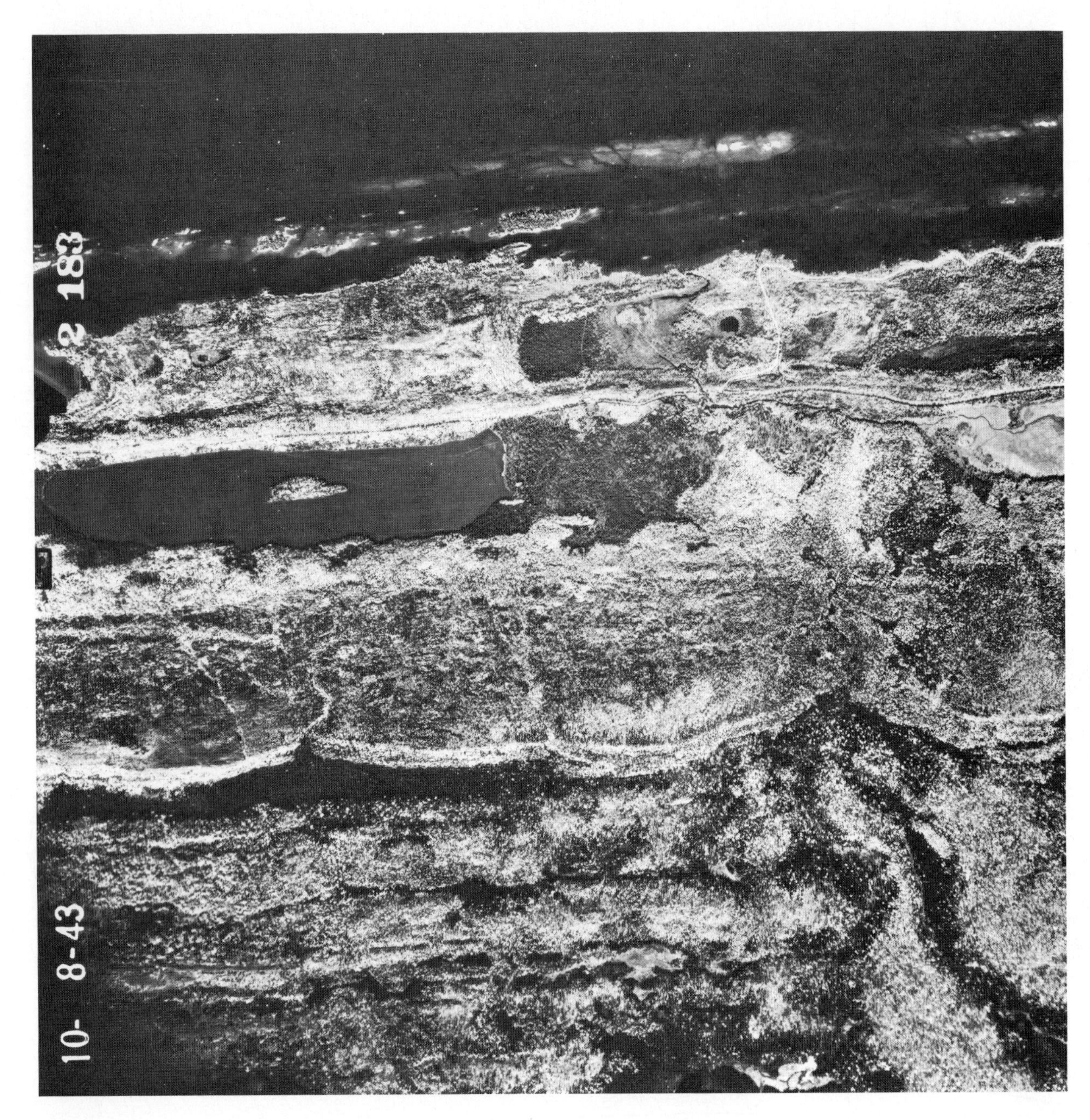

FIGURE 2.13

Aerial photograph of part of the Delaware Map in Figure 2.12. The photo is shown here for comparative purposes and will be used in a later exercise.

(Courtesy of the U.S.G.S. Photo was taken on October 8, 1943.)

Aerial Photographs and Other Imagery from Remote Sensing

Background

An aerial photograph is a picture taken from an airplane flying at altitudes as high as 60,000 feet. Practically all of the United States has been photographed aerially by federal, state, or other agencies.

Photographs taken with the axis of the camera pointed vertically down are *vertical photos* and are most useful to the geologist. Aerial photos are also used for the making of topographic maps, for highway location, military operations, mapping of soils, city planning, and for many other operations.

Other "pictures" made with special sensors mounted in earth-orbiting satellites are also useful for displaying features of the earth's surface. The false color images are perhaps the most spectacular from an aesthetic point of view, and they are quite useful in geologic interpretative work because of the broad regional relationships they show. False color images and other kinds of earth satellite imagery have been made with the imaging devices aboard the several Landsat satellites, the first of which was launched on July 23, 1973. Imagery of a variety of types including side-looking radar has been acquired from devices on board other satellites or space vehicles. The satellite data is transmitted to receiving stations on earth. These data are then processed into images of various types and are made available to users through commercial sources or agencies of the federal government such as the U.S.G.S. and National Oceanic and Atmospheric Administration (NOAA).

Space imagery results in "maps" that are very small in scale; that is, an inch on these image maps represents many miles on the ground. Thus, the Landsat maps do not have the resolution (degree of detail) that aerial photographs have. For this reason, most of the remote sensing imagery used in this manual will consist of aerial photographs. Where appropriate, a false color image will be used.

Determining Scale on Aerial Photographs

The scale of an aerial photo depends on both the focal length of the camera and the height of the airplane above ground surface. If these two factors are known, the R.F. of the photo can be determined. The photo scale can also be determined, however, by measuring known ground distances on the photo. In many parts of the coterminous United States, section lines are ideally suited for this purpose since they are normally 1 mile apart and are usually visible on the photo as roads, fences, or other human-made features. Conversion to an R.F. or graphic scale can be made in the same manner as was described under the section on topographic map scales.

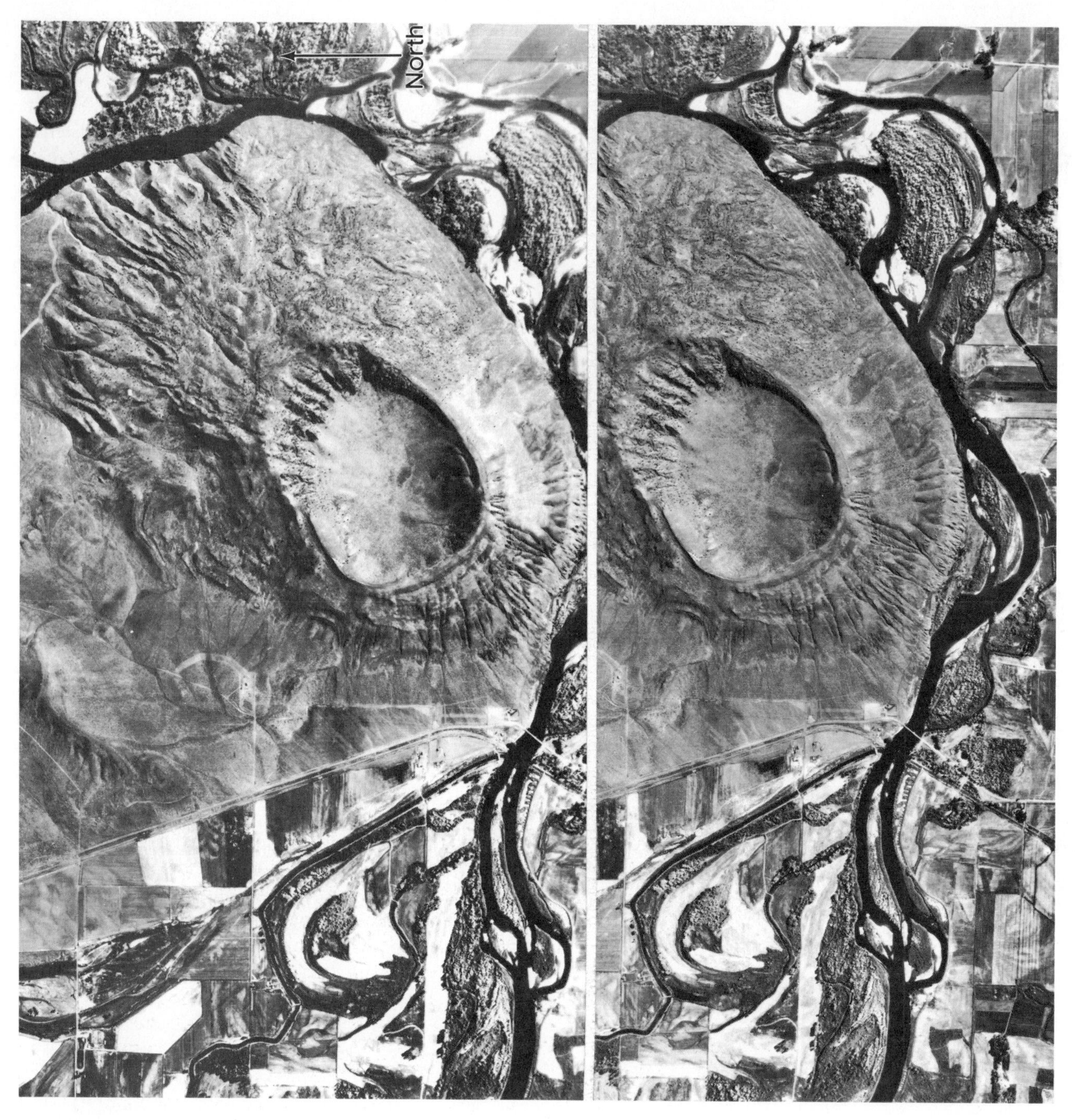

FIGURE 2.14

Stereopair of Menan Buttes, Idaho. The Snake River flows around the south side of the crater.
(Courtesy of the U.S.G.S. Photos were taken on October 8, 1950.)

STEREOSCOPIC USE OF AERIAL PHOTOGRAPHS

If two vertical photos are taken from a slightly different position and viewed through a *stereoscope,* the relief of the land becomes visible. A *stereopair* consists of two photos viewed in such a way that each of the eyes sees only one of the two photos. The brain combines the two images to form a three-dimensional view of the objects shown on the photograph.

Figure 2.14 is a stereopair for practice in using the simple lens stereoscope. Figure 2.15 shows one being used. The stereoscope is positioned over the stereopair so that the viewer's nose is directly over the line separating the two adjacent photos. If stereovision is not immediately achieved, the stereoscope should be rotated slightly around an imaginary vertical axis passing through the midpoint between the two lenses until the image viewed appears in relief.

FIGURE 2.15

Student using a lens stereoscope to view a stereopair. In actual practice, the nose is positioned in the slot between the two lenses in order to bring the eyes closer to the lenses.

On photographs in which steep slopes occur (e.g., the walls of deep canyons or the flanks of high mountains), the stereoscopic image of these features is exaggerated. That is, steeply inclined canyon walls may appear to be nearly vertical when, in fact, they are not. This distortion, however, does not present a problem under normal conditions of viewing by the beginning student.

In both Parts 3 and 4 of this manual, several exercises will require interpretation of single aerial photos and stereopairs with reference to the terrain features and geologic phenomena shown on them. In preparation for those exercises, the paragraphs that follow will provide the student with some basic information on the interpretation of aerial photographs in terms of the recognition of common natural and human-made features.

Interpretation of Aerial Photographs

Aerial photos may be used to great advantage by geologists. Stereopairs are preferable, but single photos taken under good conditions of lighting and from proper altitudes reveal exactly what the human eye sees except for the third dimension, and even this can, with practice, be approximated from single photos.

The interpretation of aerial photos is an art acquired only after considerable experience in working with photos from many different areas. However, beginning students can comprehend aerial photos amazingly well if they have a few simple instructions to follow and some basic principles to guide them.

The greatest difficulty confronting an individual looking at an aerial photo for the first time is recognition of familiar features that are seen every day on the ground but become mysterious objects when seen from the air. It is necessary, therefore, to acquire the ability to recognize certain common features before one can expect to use an aerial photo as a geologic tool. Some of these common features are described in the paragraphs that follow and are illustrated in figure 2.16.

Vegetation

Vegetational cover accounts for a great many differences in pattern and shades of gray tone on aerial photos. Heavily forested areas are usually medium to dark gray, whereas grasslands show up in the lighter tones of gray. Planted field crops are extremely varied in tone, depending not only on the kind of crop but also on the stage of growth. Cultivated fields are usually rectangular in shape and appear either in dark gray or light gray, depending upon whether the fields have just been plowed or whether the crop is already in and growing.

Soil and Rock

Soil texture controls the soil moisture, which in turn controls the appearance of the soil on aerial photos. Wet clayey soils have a much darker shade of gray than do the dry sandy soils, which usually show up as light gray to almost white in arid regions. Different tones of gray that show up in the same field are due to different degrees of wetness of the soil, a condition usually related to topographically low (wet) and high (dry) areas.

Where bedrock crops out at the surface, aerial photos reveal differences due to lithology, texture, mineral composition, and structure of the rocks. The beginning student cannot always discern between a sandstone and a limestone, for example, but should have less trouble recognizing the differences among the major rock groups (i.e., igneous, sedimentary, and metamorphic). Vegetational patterns on aerial photos commonly reflect the underlying bedrock, and this is helpful in tracing out a single rock unit of the photo. On the other hand, if a thick cover of residual material such as soil or talus, or a cover of transported material such as glacial drift, alluvium, or eolian sands cover the area, then all bedrock features may be partially or totally obscured.

Aerial photos used in connection with field investigations on the ground are among the most valuable tools of the professional geologist. When direct field examination is not feasible, aerial photos provide even better information than one could acquire by flying over the area in person, especially if the photographs are available for stereoscopic study. For the student of elementary physical geology, aerial photographs are useful in that they show geologic features of the earth's surface that otherwise would be difficult to describe or impossible to illustrate.

A. Sedimentary rocks uplifted to form dome (egg-shaped feature) in a semi-arid climatic zone. Stream valley with deciduous trees in valley bottom (black). Wyoming. Scale, 1:80,000.

B. Deeply incised river in flat lying sedimentary rocks. Badlands topography on either side of the river. Colorado. Scale, 1:80,000.

C. Small town in Midwest. Various cultural features shown. Deciduous vegetation in town and along river. Missouri. Scale, 1:20,000.

D. Agricultural field patterns (light and dark variegated areas) with undrained depressions forming ponds (white due to reflection of sun off water surface). Contour plowing evident. Texas. Scale, 1:63,360.

E. Crystalline rocks with sparse coniferous vegetation. River in deeply cut valley. Lake (black) at margin of photo. Joint pattern in rocks evident. Wyoming. Scale, 1:60,300.

F. Glacial moraine from continental glacier. Kame and kettle topography with numerous kettle lakes (black). Rectangular field patterns in variegated colors. North Dakota. Scale, 1:60,000.

FIGURE 2.16

Some examples of features visible on aerial photographs.

Landsat False Color Images

The normal human eye can perceive a continuous spectrum from blue and green to yellow, orange, and red. Neither ultraviolet nor infrared radiation is visible to humans. The imaging technology used on the Landsat remote sensors records infrared radiation that, when combined with other wavelengths in the processing of the imagery data, results in a colored image in which the true colors are replaced by other colors. Hence the term, *false color image.*

In false color images, green vegetation shows up as various shades of red; deserts and other nonvegetated tracts are light gray to bluish gray; cities and large metropolitan areas are dark gray; and clear water areas are commonly black. Water bodies containing silt and other suspended sediments appear in various shades of light blue. Because vegetational patterns reflect to some extent the underlying soil and rock types, the various shades and hues of red and other colors on a false color image define the different kinds of soil and rock types.

Landsat views do not cover a rectangular area due to the fact that the earth rotates as the satellite passes overhead. This phenomenon results in a rhombohedral "picture" such as the image shown in figure 2.17. North is generally toward the top margin of the image, but a true north-south line must be determined independently of the image margins. Where visible, agricultural fields are useful for this purpose because their boundaries are commonly oriented in north-south and east-west directions. If these are not present, other features such as a coastline or a major river can be compared with a map showing the same features to determine true north.

Figure 2.17 is a false color image of part of the Atlantic coast of the United States showing Long Island, the Hudson River, and a part of New Jersey. The Atlantic Ocean appears black, and the light-colored patches in New Jersey are cultivated fields. The brownish area in the southern part of the image is the forested coastal plain of New Jersey, and the dark area in the extreme northwest corner is part of the northern Appalachian Mountains. New York City and Newark, New Jersey, at the mouth of the Hudson River, are dark blue-gray. The light color of the narrow strips of coastal beaches off the southern shore of Long Island and off the coast of New Jersey reflects the fact that these features are sandy and generally lack heavy vegetation.

References

Avery, T. E., and Graydon, L. B. 1985. *Interpretation of aerial photographs.* 4th ed. Minneapolis: Burgess Publishing Co. 554 pp.

Colwell, R. N., ed. 1983. *Manual of remote sensing.* 2d ed. Falls Church, Virginia: American Society of Photogrammetry and Remote Sensing. 2,724 pp.

Drury, S. A. 1987. *Image interpretation in geology.* London: Allen and Unwin. 243 pp.

Hyatt, E. 1988. *Keyguide to information sources in remote sensing.* London: Mansell Publishing Ltd. 274 pp.

Sabins, F. F. 1986. *Remote sensing: principles and interpretation.* 2d ed. Oxford: W. H. Freeman and Co. 592 pp.

Williams, R. S., and Carter, W. D. eds. 1976. *ERTS-1, a new window on our planet.* U.S.G.S. Professional Paper 929, Washington, D.C. 362 pp.

FIGURE 2.17

False color image of the Long Island—New Jersey area made from Landsat 2 on October 21, 1975. The long narrow strip of sand beach off the southern shore of Long Island is about 80 miles long.

(NASA ERTS image E-2272–14543, U.S.G.S. EROS Data Center, Sioux Falls, South Dakota 57198.)

Name Section Date

EXERCISE 11
INTRODUCTION TO AERIAL PHOTOGRAPH INTERPRETATION

1. Figure 2.18 is an aerial photograph of standard size. Identify the following features on the photograph. Where a topographic map symbol is available for the feature (refer to fig. 2.5), draw the symbol directly on the photograph of figure 2.18 at the place where the feature occurs. If no symbol exists, draw the outline of the feature on the map and label it accordingly. Use a red pencil in all cases except for water features, which should be shown in blue.
   (a) Major road
   (b) Railroad track
   (c) River
   (d) Landing strip
   (e) Football field and track
   (f) Small stream valley
   (g) Bridge over river
   (h) Overpass
   (i) Golf course
2. The southeast corner of the golf course is also the SE corner of Section 12, T. 92 N., R. 52 W. The "T" intersection of the road at the south edge of the golf course about 4 inches to the west (left) on the photograph is at the SW corner of the SE¼ of the SE¼ of Section 11, T. 92 N., R. 52 W. Determine the R.F. and verbal scale of the photograph.

   R.F. ______________________

   Verbal Scale ______________________

   ____
3. Figure 2.19 is a stereopair. Examine it with a stereoscope and answer the following questions while completing the instructions.
   (a) Does the major stream channel contain any signs of the presence of water?

   ____

   ____

   (b) Trace the drainage lines in blue pencil on the right-hand photograph using the proper symbol from figure 2.5.
   (c) What is the dominant vegetation of the area?

   ____

   ____

   (d) Are any human-made features visible on the stereopair? ______________________

   ____

   (e) The area covered by the stereopair shows two relatively flat upland surfaces away from the stream channel, each of which lies at a general elevation that differs from the other. Draw the boundaries of these areas on one photograph while viewing the stereopair with a stereoscope. Use a red pencil. Mark the lower-lying area with the number *1,* the higher area with the number *2.* Extend these boundaries on the rest of the photograph.
4. Figure 2.13 covers part of the Delaware map of figure 2.12. With red pencil, label the following features with the lowercase letter that appears before each name:
   (a) Lake Superior
   (b) Lake Bailey
   (c) Agate Harbor
   (d) Small lake in section 35
   (e) Swamp area south of Mt. Lookout
   (f) Mt. Lookout
   (g) Trace the roads in sections 34, 35, and 36 that are shown by a dashed red line on the Delaware map.
5. Trace the drainage system from the Delaware map onto the aerial photo of figure 2.13. Use a blue pencil and label each creek or river that has a name.
6. Determine the scale of the aerial photo in figure 2.13.

7. Figure 2.3 is an aerial photograph of an area in central Iowa. Section 6 of T. 90 N., R. 37 W. is labeled on the photograph.
   (a) Using the information provided, determine the scale of the photograph. ____________________

   ____

   (b) Draw the township and range lines on the photograph using red pencil. Draw in the section lines with black pencil. Number all sections in black pencil. Along the photograph margins record the township and range designations in red pencil using the conventional system shown in figure 2.2.
   (c) Note the relationship between the section lines and the road pattern. Compare this relationship with that on figure 2.18. Suggest possible reasons for the differences.

   ____

   ____

   ____

   ____

   ____

   (d) A railroad track crosses the area from southeast to northwest. Use the appropriate map symbol to show this feature on the photograph. (Draw directly on the photograph with black pencil.)
   (e) The rectangular areas covering most of the area are croplands. They appear in various shades of gray to nearly black on the photo. Explain the differences.

   ____

   ____

   ____

   ____

8. The scale of a vertical photograph depends on the focal length of the camera used and the height of the airplane taking the photo above the ground. The R.F. is the focal length of the camera divided by the height of the plane above the ground. Determine the scale of a photograph taken by a camera with a 9-inch focal length from an airplane flown at a height of 35,000 feet above the ground. (Be careful about units.)

SHOW YOUR CALCULATIONS:

FIGURE 2.18

Aerial photograph of area in South Dakota.

(Photo No. VE-1JJ; taken on June 19, 1968.)

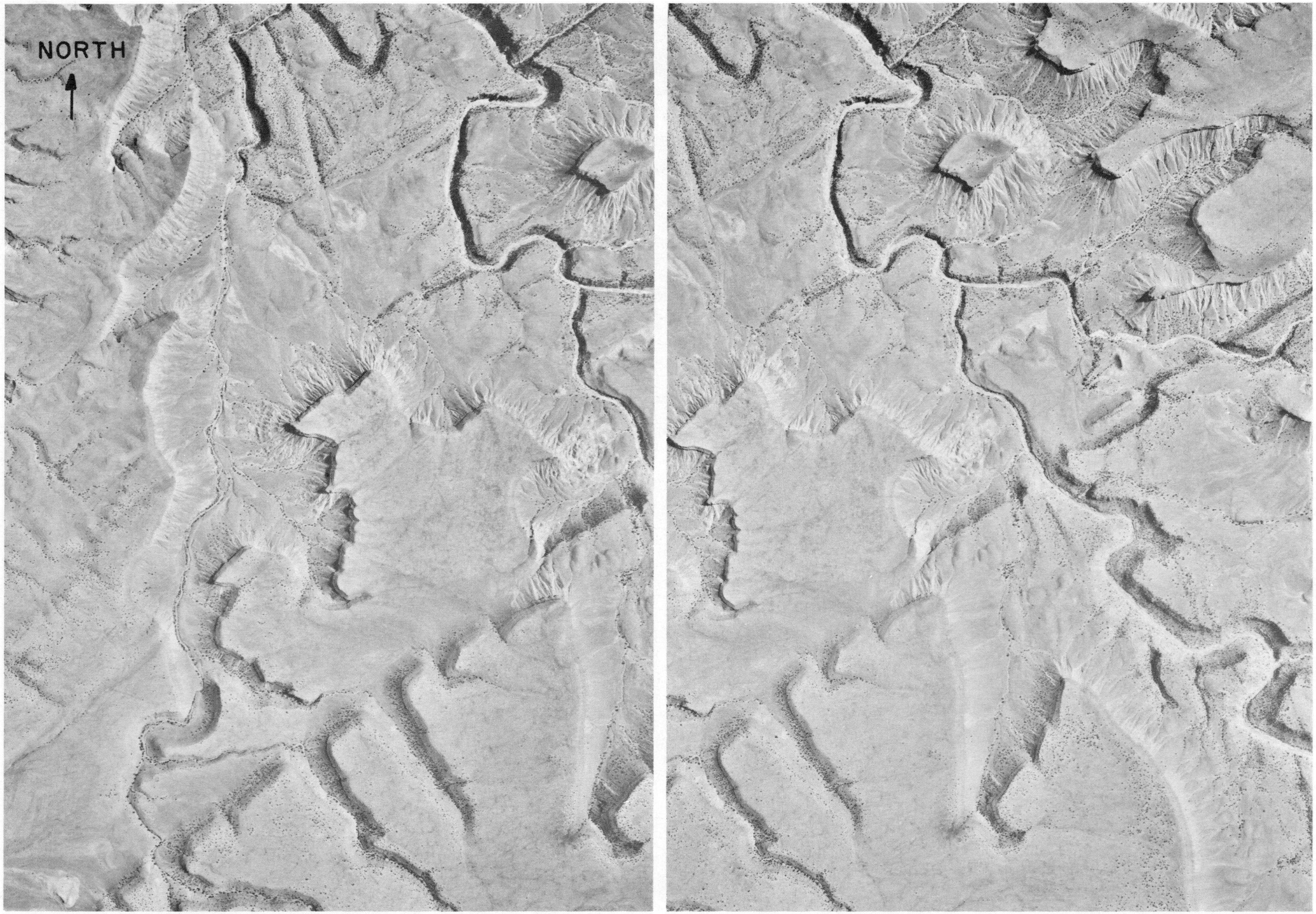

FIGURE 2.19

Aerial Photograph stereopair, Utah, 1956. Scale, 1:20,000. (Photograph numbers GS-RR-17–43 and GS-RR-17–44.)

PART THREE

GEOLOGIC INTERPRETATION OF TOPOGRAPHIC MAPS, AERIAL PHOTOGRAPHS, AND EARTH SATELLITE IMAGES

BACKGROUND

In Part Two of this manual you learned that the configuration of a landform is expressed on a topographic map by contour lines and is revealed on aerial photographs when two overlapping photos are viewed stereoscopically. Topographic maps and stereopairs can thus be used for the study and analysis of terrain features in terms of the geologic processes that produced them. Images from earth-orbiting satellites are additional tools useful in the study of landforms.

Every geologic process leaves some imprint on the part of the earth's surface over which it has been operative. These processes include the work of the major geologic agents such as wind, groundwater, running water, glaciers, waves, and volcanism. Each of these agents leaves its mark on the landscape in the form of one or more characteristic landforms.

The association of geologic agents with the origin of various landforms is a subdivision of geology called *geomorphology.* Geomorphologists have systematized the relationship of geologic processes to topographic forms into a body of knowledge that can be used in deciphering the origin of topographic features shown on a topographic map or seen in a stereopair. The body of knowledge that deals with the origin of landforms is presented in all basic textbooks that deal with physical geology. Figure 3.6 is an artist's rendition of the landforms of North America.

It is assumed that the users of this manual will have become acquainted with the different geologic processes and their related landforms through reading appropriate chapters in a textbook on physical geology and by listening to lectures in which the origin of landforms is presented. This is prerequisite to the understanding and successful completion of the exercises that are presented in this part of the manual.

GENERAL INSTRUCTIONS

The purpose of Exercises 12 through 19 is to acquaint you with a variety of landforms and geologic principles associated with the geologic agents of wind, groundwater, running water, glaciers, waves, and volcanism. Topographic maps and profiles, aerial photographs (some of which are in the form of stereopairs) and other pertinent maps, satellite images, diagrams, and data are provided in Exercises 12 through 19 as the basic tools for learning the association between landform and geologic agent, or to establish a specific geologic principle.

The title of each exercise identifies the geologic agent that will be under consideration for that particular exercise. It is assumed that you are thoroughly conversant with and have a good understanding of the material presented in Part 2 of this manual—on topographic maps, satellite images, and aerial photographs, including map and photo scales, contour lines, map symbols, and topographic profiles. The terms associated with the geologic processes covered in this part of the manual generally are defined in the background material for each exercise in which they are used. *However, it is good practice to bring your textbook to the laboratory as a reference for unfamiliar or forgotten terms that crop up in the exercises.*

All maps and photographs needed for completion of the exercises are included in the manual. Before proceeding with the questions or problems based on maps or

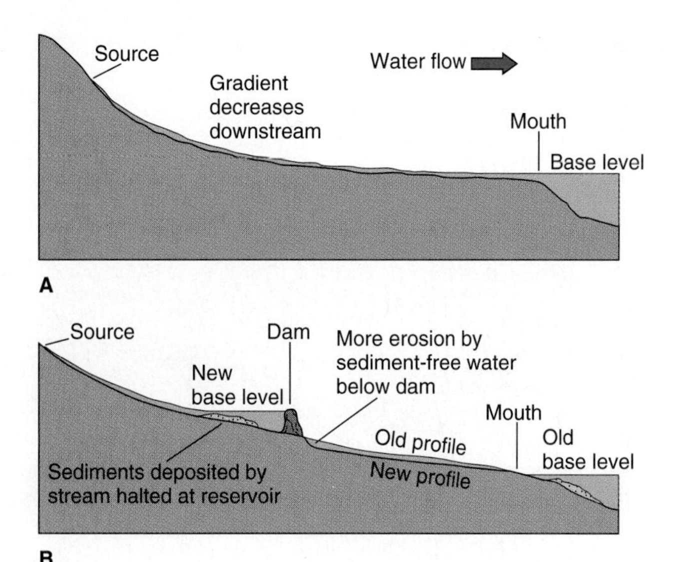

FIGURE 3.1

(*A*) Longitudinal profile of a river showing a gradual decrease in gradient downstream. Base level is the lowest level to which a stream can erode its bed. (*B*) Longitudinal river profile that has been interrupted by a dam. The old pre-dam base level has been replaced by a new one.

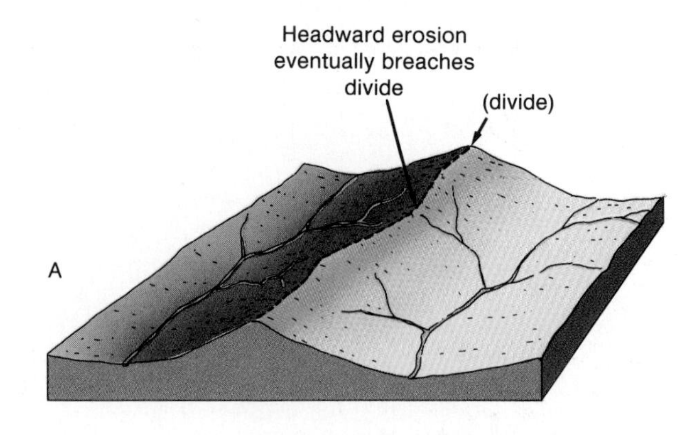

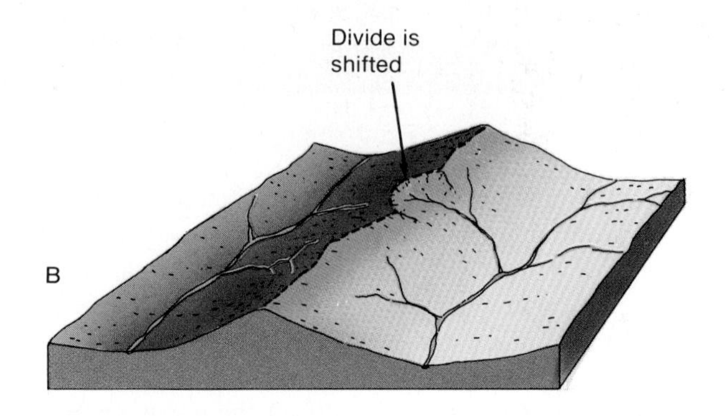

FIGURE 3.2

(*A*) A divide between two opposing drainage systems is attacked by headward erosion. (*B*) The divide is shifted toward the drainage system whose streams have the smaller gradient of the two.

photographs, note the scale and contour interval on the topographic maps and the scale of the stereopairs. Some exercises require you to draw on the maps or photos with ordinary lead pencil or colored pencils. In these cases, make your initial lines very light so that if erasure is necessary, it can be done easily. Also, use sharp pencils to ensure accuracy, especially where the drawing of a topographic profile is a requirement of the exercise. Some students who suffer from eyestrain after prolonged study of maps may find an inexpensive magnifying glass helpful for completing the map exercises. The questions for each exercise should be answered in the order given because they are arranged in a more or less logical sequence.

Geologic Work of Running Water

Stream Gradient and Base Level

Landforms produced by running water are produced by stream erosion, stream deposition, or a combination of erosion and deposition. The *gradient* of a stream is the slope of the stream bed, or surface of the stream in larger rivers, along its course. A *longitudinal profile* of a stream shows that the gradient decreases in a downstream direction (fig. 3.1A). A stream tends to erode its channel bottom or bed in the upper or headward reaches and deposit sediment or erode its channel walls in the lower reaches.

Base level is the lowest level to which a stream can erode its bed. The base level for streams that flow to the ocean is sea level, but other temporary base levels may exist along the stream course. These include lakes resulting from natural processes, such as landslides or lava flows damming the stream, or the dam may be a human-made feature, such as Hoover Dam on the Colorado River or smaller dams on local rivers (fig. 3.1B).

The surface of a lake behind a dam constitutes a new base level for the segment of the stream above the dam. Sediment is deposited where the stream enters the lake because the stream gradient has been reduced. Below the dam, a new profile is formed because the water discharged through the gates of the dam is relatively free of sediment, thereby allowing the stream to erode its bed to a deeper level than before the dam was built.

Stream Gradients and Drainage Divides

A stream together with its tributaries is called a *drainage system.* Contiguous drainage systems are separated by a *drainage divide,* an imaginary line connecting points of highest elevation between the two systems (fig. 3.2A).

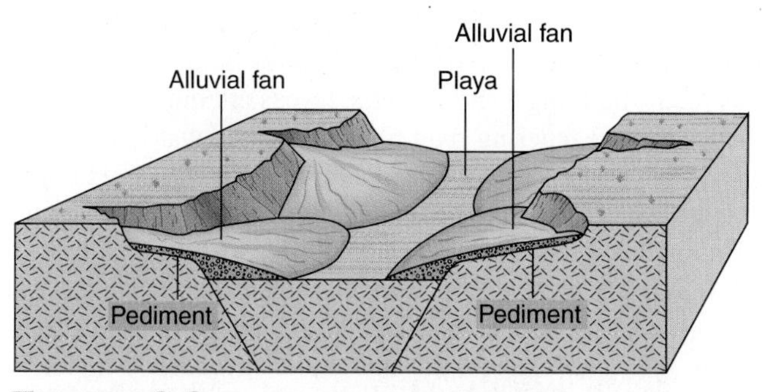

FIGURE 3.3

Diagram showing relationship between alluvial fans and a pediment.

From Charles C. Plummer and David McGeary, Physical Geology, 8th edition.

The erosive power of a stream is a function of the stream gradient. Streams with higher gradients are more potent in eroding their beds than streams with lower gradients. If adjacent drainage systems have comparable gradients in their upper reaches, the divide between the two systems will remain more or less constant. If, however, the gradients of the streams in the two systems differ appreciably, the divide will move over time toward the system with the smaller gradient (fig. 3.2B).

Alluvial Fans and Pediments

These two landforms have similar but not identical topographic expressions. An *alluvial fan* is a depositional feature produced where the gradient of a stream changes from steep to shallow as it emerges from a mountainous terrain (fig. 3.3). At that point the coarse sediments carried by the fast-flowing stream in the mountain are deposited in the form of a fan-shaped apron at the mountain front where the gradient decreases abruptly.

A *pediment* (fig. 3.3), on the other hand, is a gently sloping *erosional surface* covered with a veneer of coarse sediment. Pediments are common in arid regions. They form as the weathered mountain front recedes by attack from heavy but infrequent rainfalls. Below the coarse sedimentary cover in transit lies the planed-off bedrock that was formerly part of the mountainous terrain. One description of a pediment is that it is the end product of a mountain range consumed by weathering and stream erosion. A pediment may contain remnants of the mountainous terrain standing as isolated bedrock knobs above the pediment surface.

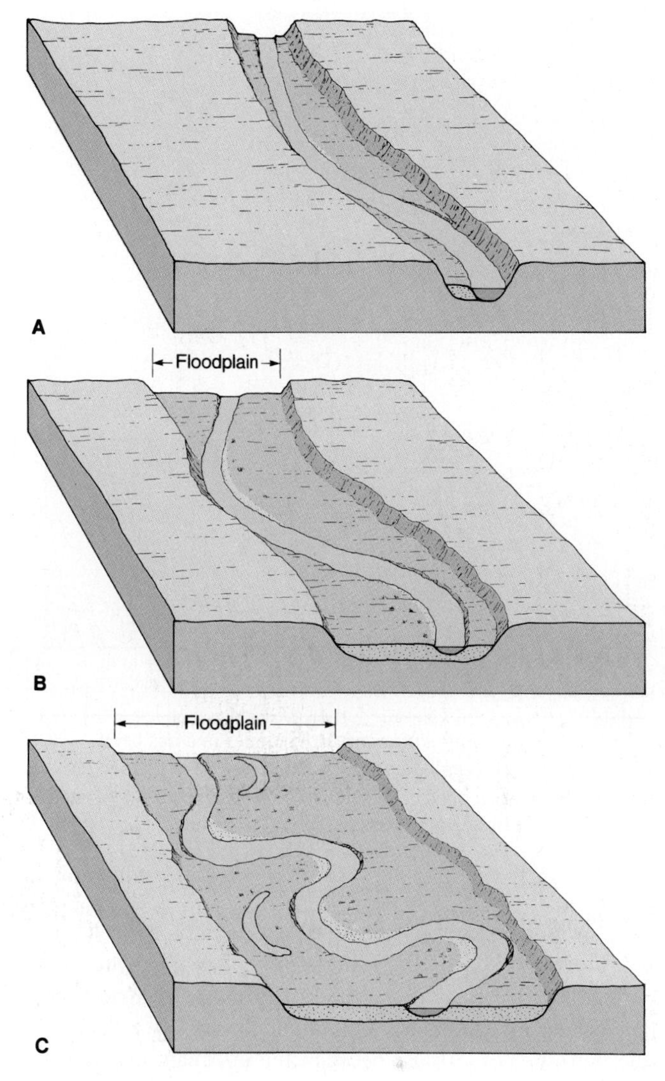

FIGURE 3.4

The evolution of a floodplain. (*A*) River widens the valley floor by lateral erosion. (*B*) Continued lateral erosion widens the valley floor further, and the river course becomes more meandering. (*C*) The river's course develops meander loops, which become oxbow lakes.

(From Carla W. Montgomery, Physical Geology, 3d ed.

Meandering Rivers and Oxbow Lakes

In the lower reaches of a stream's course where the gradient is low, the stream's erosive action is directed toward the channel walls (fig. 3.4A). This process of *lateral erosion* produces a flat valley floor called a *floodplain* (fig. 3.4B). A floodplain is not produced by flood stages of the river but rather by constant shifting of the stream channel through lateral erosion across the valley floor. A floodplain does become inundated when the stream channel is unable to accommodate the volume of spring runoff or of heavy rains in the drainage basin at any time of the year.

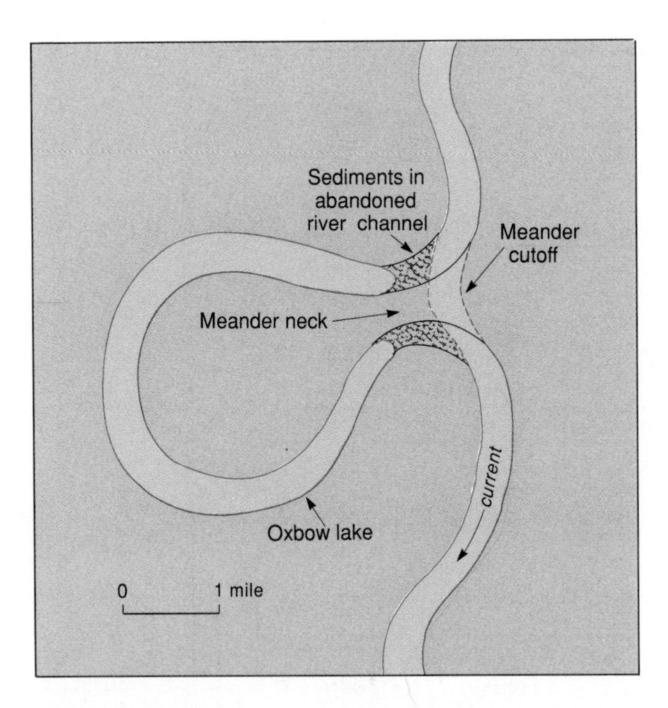

FIGURE 3.5

Map diagram showing a former river meander that has been severed from the river by a meander cutoff that flows across the meander neck. The ends of the old meander have been filled with sediment, thereby forming a closed depression called an oxbow lake.

The circuitous course of a stream flowing across its floodplain is called a *meandering course,* and the individual bends are called *meanders* (fig. 3.4C). When a meander loop is isolated from the main channel by erosion of the narrow neck of land between the upstream and downstream segments of the meander loop, an *oxbow lake* is formed (fig. 3.5). With passage of time, an oxbow lake gradually becomes filled with sediment brought in by flood waters and eventually is reduced to a swampy scar on the floodplain as testimony to its former existence as an active channel of the river.

Floodplains are attractive sites for agriculture and land development. To protect structures built on them such as roads, buildings, and airports, levees are built along both sides of a meandering river to contain flood waters.

The lower Mississippi River is a classic example of a meandering stream flowing across a broad floodplain. Hundreds of miles of levees have been constructed by the U.S. Army Corps of Engineers to contain flood waters. Additionally, the Corps of Engineers has cut off meanders from the main channel through the excavation of channels across meander necks. These artificial *cutoffs* shortened the river course in deference to the barge traffic plying the river with cargoes of various sorts.

The Mississippi River forms the boundary between many states, from Minnesota to its mouth in Louisiana where it enters the Gulf of Mexico. The states of Mississippi and Arkansas generally lie east and west of the river, respectively. Some small segments of each of these states, however, lie on the *opposite* side of the river. The reason for this is based on the meandering nature of the river and the need to fix the boundary so that the shifting course of the river does not also shift the state boundary. To reduce interstate conflicts that might arise through a constantly shifting boundary, the United States Supreme Court ruled in 1820 that where a stream forms the boundary between states, and the channel of the stream changes by the "natural and gradual processes known as erosion and accretion, the boundary follows the varying course of the stream; while if the stream from any cause, natural or artificial, suddenly leaves its old bed and forms a new one, by the process known as avulsion, the resulting change of channel works no change of boundary, which remains in the middle of the old channel, although no water may be flowing in it."*

**Van Zandt, Franklin K. 1976. Boundaries of the United States and the Several States. U.S.G.S. Professional Paper 909, p. 4.*

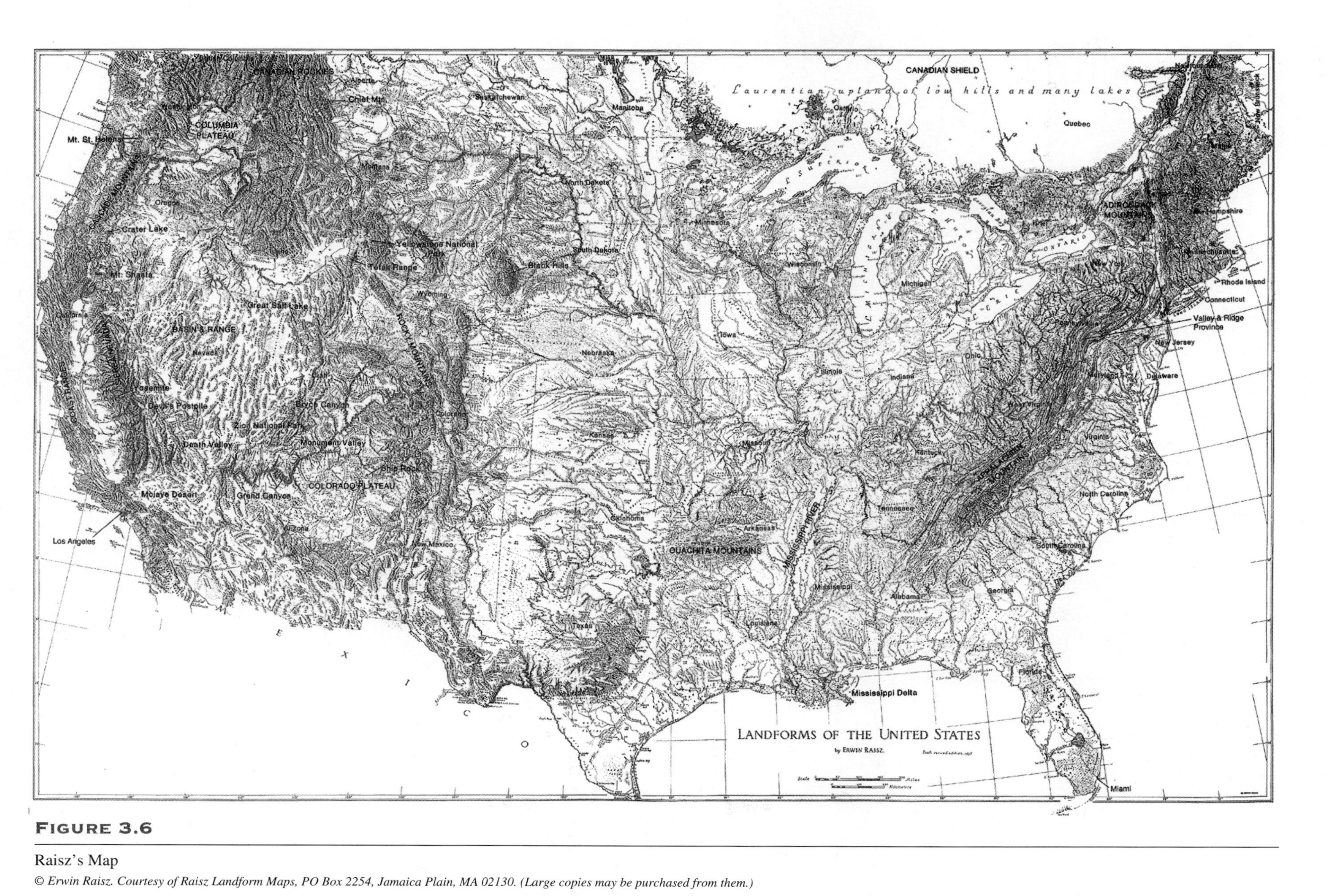

FIGURE 3.6

Raisz's Map

© Erwin Raisz. Courtesy of Raisz Landform Maps, PO Box 2254, Jamaica Plain, MA 02130. (Large copies may be purchased from them.)

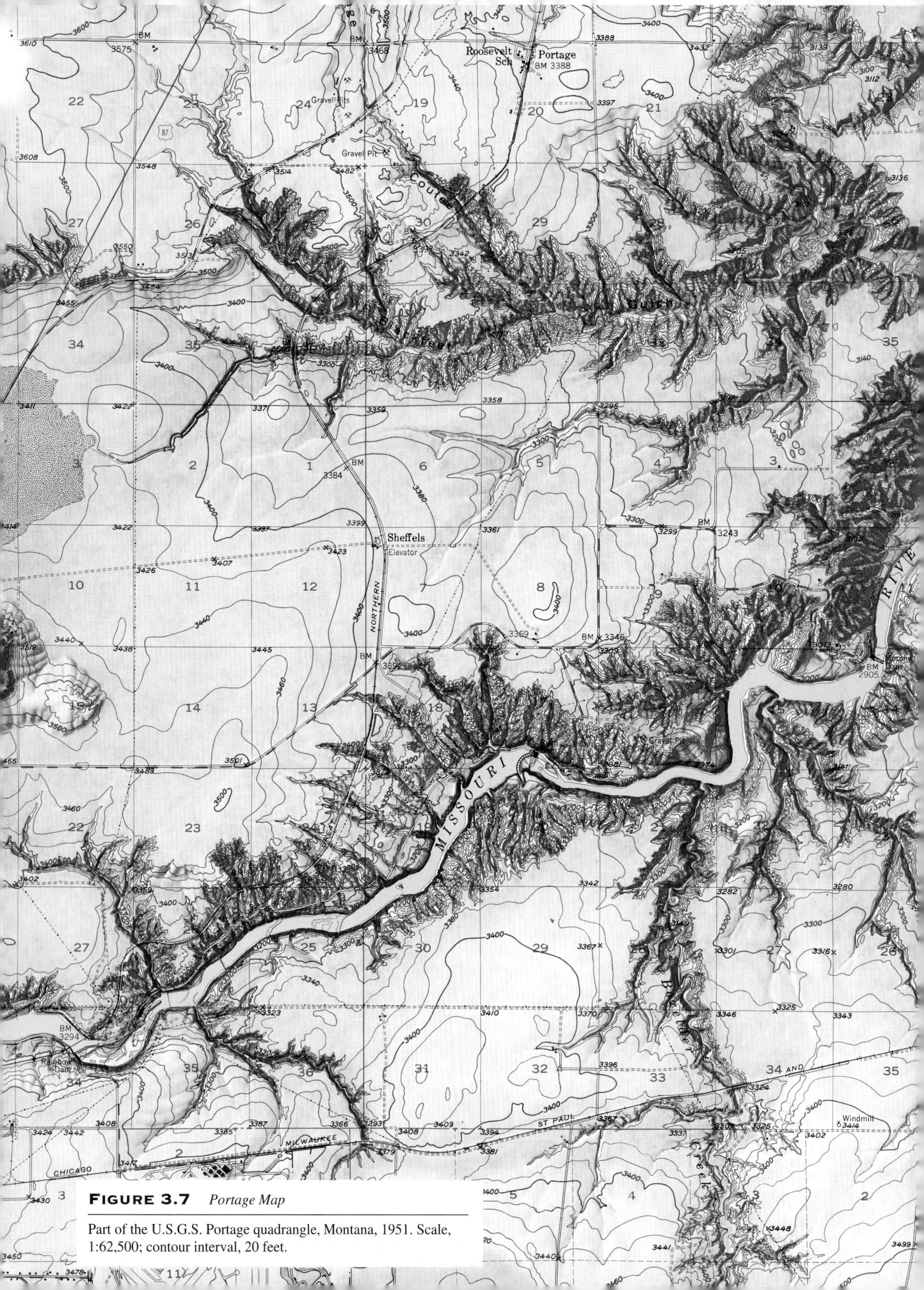

FIGURE 3.7 *Portage Map*

Part of the U.S.G.S. Portage quadrangle, Montana, 1951. Scale, 1:62,500; contour interval, 20 feet.

FIGURE 3.8 *Promontory Butte Map*

Part of the U.S.G.S. Promontory Butte quadrangle, Arizona, 1951. Scale, 1:62,500; contour interval, 50 feet.

FIGURE 3.9 *Antelope Peak Map*

Part of the U.S.G.S. Antelope Peak quadrangle, Arizona, 1963. Scale, 1:62,500; contour interval, 25 feet.

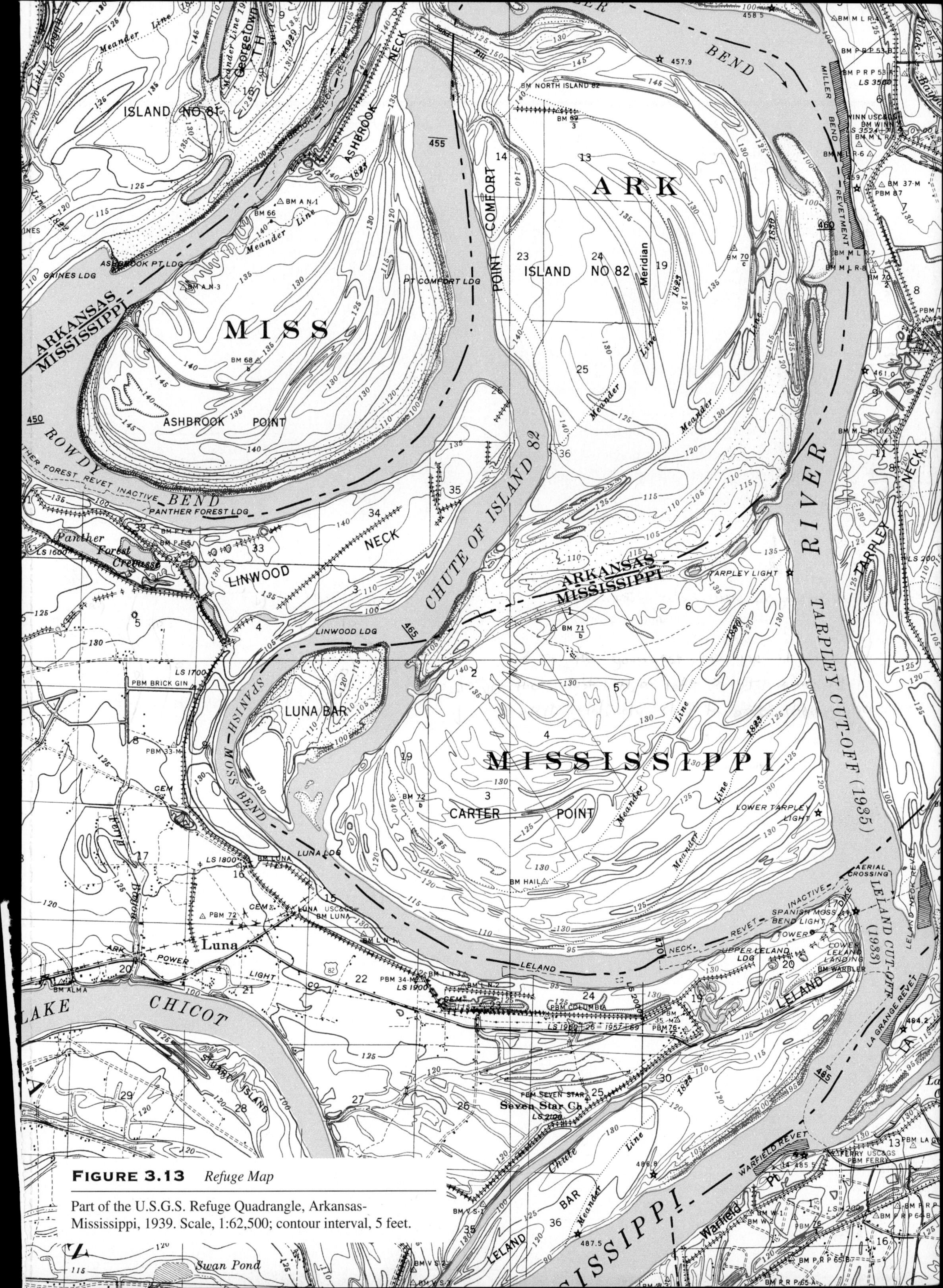

FIGURE 3.13 *Refuge Map*

Part of the U.S.G.S. Refuge Quadrangle, Arkansas-Mississippi, 1939. Scale, 1:62,500; contour interval, 5 feet.

EXERCISE 12E CONTINUED

3. Figure 3.14 shows a segment of the Mississippi River along the border between the states of Arkansas and Mississippi. The course of the modern river is not coincident with the boundary line between the two states, which was fixed by Congress before the modern course of the river was established.

   We will refer to the course of the Mississippi River shown on figure 3.14 as the "Modern Course," and the course of the river when the Arkansas-Mississippi boundary was set as the "Boundary Course." From the northern to southern boundaries of the Greenwood map, the Modern Course is 65 miles and the Boundary Course is 136 miles.

   (a) Express as a percent the amount of shortening of the Modern Course compared to the Boundary Course.

   ____
   ____
   ____
   ____
   ____

   (b) What is the impact of this shortening on the *competence* and *capacity* of the river?

   ____
   ____
   ____
   ____

4. Draw the boundaries of the Refuge map on the Greenwood map. The Refuge map was published in 1939 and the Greenwood map in 1979. Compare the Refuge map with its corresponding area on the Greenwood map. Describe the changes in former or modern channels that have occurred in this 40-year period with particular reference to the following:

   (a) Areas around Linwood Neck, Luna Bar, and Point Comfort.

   (b) The course of the Tarpley Cut-off with respect to the dike on its east side.

   (a) ______________________________
   ____
   ____
   ____
   ____

   (b) ______________________________
   ____
   ____
   ____
   ____

FIGURE 3.14 *Greenwood Map*

Part of the U.S.G.S. Greenwood, Miss.-Ark.-La. Map, 1953, revised in 1979. Scale, 1:250,000; contour interval, 50 feet with supplementary contours at 25-foot intervals.

Web connections

http://www.cst.cmich.edu/units/gel/LINKS/gellinks.htm

Resources for earth science. Links are organized around topics typically taught in introductory earth science courses. Selected on the basis of image quality and ease of use.

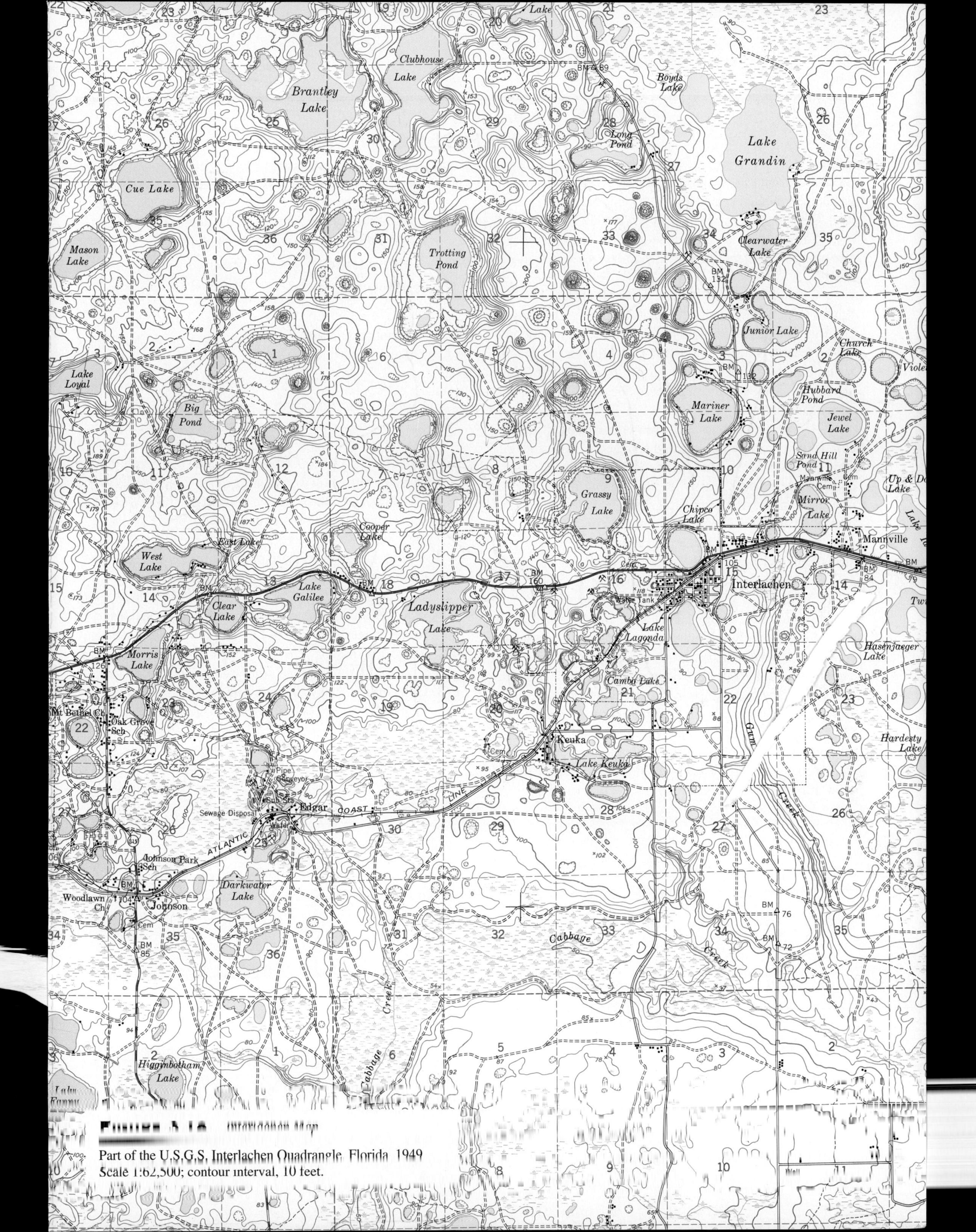

Part of the U.S.G.S. Interlachen Quadrangle, Florida, 1949
Scale 1:62,500; contour interval, 10 feet.

EXERCISE 14B
SINKHOLES

Geologic events of a catastrophic nature are usually relegated to earthquakes, landslides, volcanic eruptions, and the like. The subsidence of a part of the earth's surface, while not as devastating as the foregoing, is nonetheless catastrophic when it occurs in the course of a few hours or days.

Such an event occurred on May 8–9, 1981, in Winter Park, a town in central Florida. Figure 3.19 shows a large collapsed zone in suburban Winter Park. The collapsed feature is a sinkhole about 115 × 100 meters in horizontal dimensions. The water surface in the bottom of the depression is about 13 meters below the surrounding undisturbed land surface.

The geology of the area consists of limestone bedrock overlain by uncemented sediments (i.e., sand, silt, clay, or a mixture of these). The limestone contains large openings or caverns formed by the gradual dissolving action of groundwater percolating along joints and bedding planes. As the initial voids grow larger with the passage of time, they coalesce into a large cavernous opening. Gradually, the material from the *base* of the overlying sediments moves downward into the cavern thereby leaving a large cavity in the lower part of the unconsolidated layer, but the upper part of this layer remains undisturbed. Eventually, however, the overlying sediments collapse suddenly to fill the empty space at the bottom of the sedimentary layer, and a conical sinkhole such as the one in figure 3.19 is formed.

When such an event occurs in an urbanized area, considerable property damage results. Human-made structures collapse, sewer lines are ruptured, and underground utility lines (gas, pipelines, electric cables, telephone lines, etc.) are left hanging or are severed. Study figure 3.19 and answer the following questions.

1. Identify the following human-made features that have been damaged or destroyed by the collapse. (Write the letter of the feature directly on the photo, e.g. "a" for paved street, "b" for swimming pool, etc.)
   (a) Paved street
   (b) Swimming pool
   (c) Parking lot
   (d) Unbroken utility line
   (e) Broken utility line
   (f) A van or recreation-type vehicle
2. Figure 3.20 is a sketch of the probable cross-section of the geologic conditions that prevailed long before the Winter Park Sinkhole was formed. It depicts a network (not to scale) of voids (shown in black) in the limestone caused by groundwater solution along joints and bedding planes. (The outline of the darker blue area shows the incipient topographic profile of the Winter Park Sinkhole.) Figure 3.21 shows the geological conditions that prevailed immediately after the collapse. During the geologic time that lapsed between the conditions depicted on Figures 3.20 and 3.21, the limestone continued to dissolve until a large cavern formed beneath the site of the future sinkhole. When certain conditions were achieved, the collapse occurred.

   A description of the process of sinkhole formation in unconsolidated sediments resting on a cavernous limestone bedrock is described on page 129. Read it carefully and then complete the geologic cross section of figure 3.22 showing the conditions that existed *immediately prior* to the collapse that formed the Winter Park Sinkhole (assume that the base of the large limestone cavern is below the bottom of the diagram).

FIGURE 3.19

Photograph of Winter Park Sinkhole taken on May 13, 1981 by Rick Duerling.
(Courtesy of Florida Bureau of Geology, Tallahassee, Florida.)

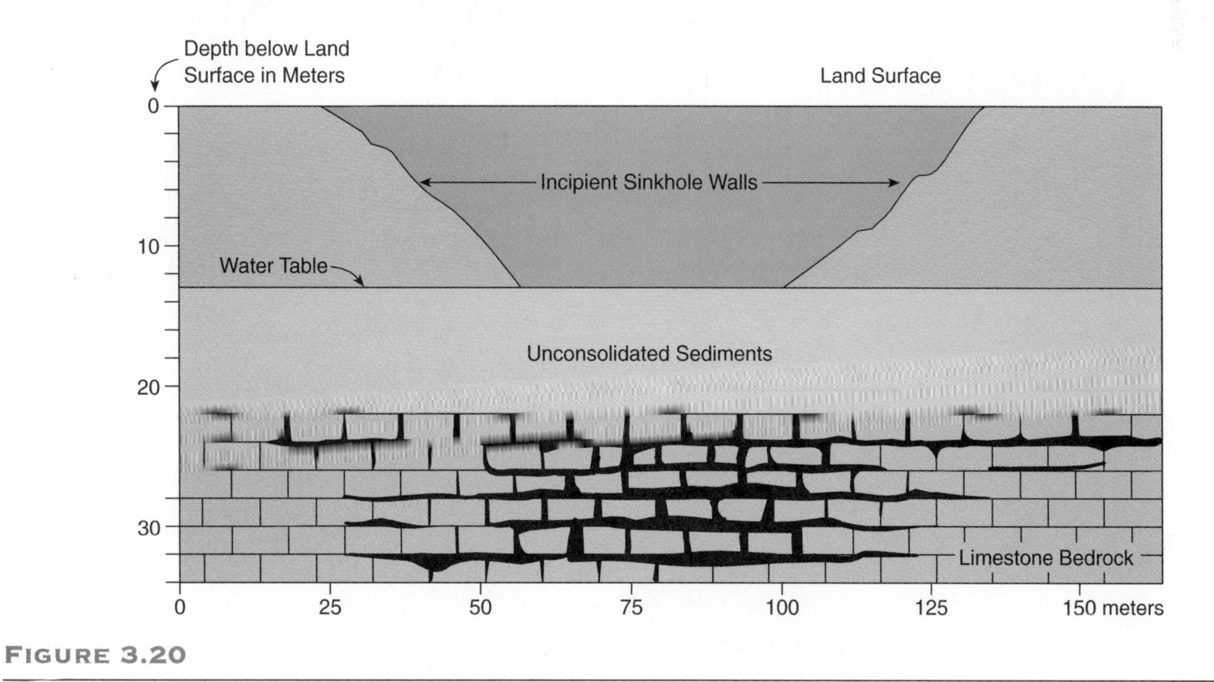

FIGURE 3.20

Geologic cross section of Winter Park Sinkhole area long before collapse occurred.

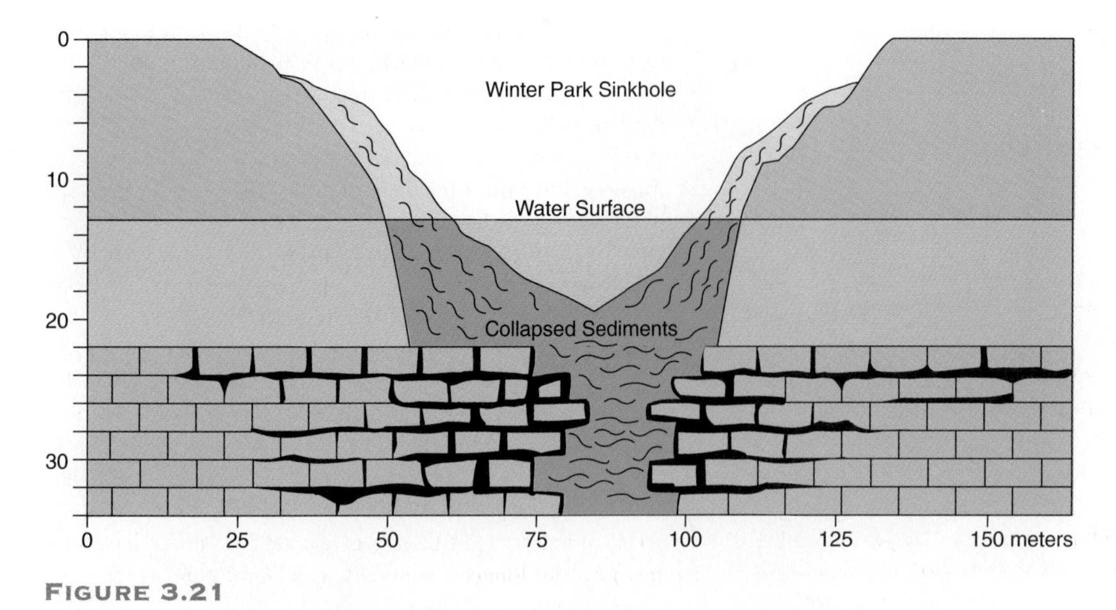

FIGURE 3.21

Geologic cross section of Winter Park Sinkhole *immediately after* sinkhole was formed.

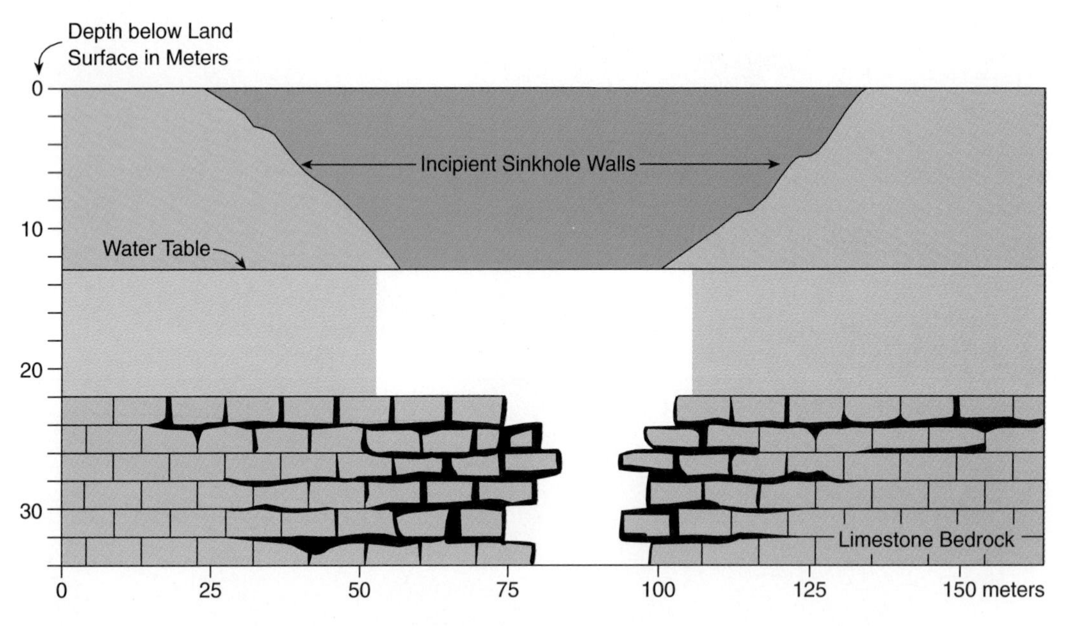

FIGURE 3.22

Incomplete geologic cross section of Winter Park Sinkhole *immediately before* collapse. (See Exercise 14B, question 2, for instructions.)

Sinkhole Formation

A Description of The Process of Sinkhole Formation In Unconsolidated Sediments Overlying Cavernous Limestone.

(From Foose, R. M., 1981, Sinking can be slow or fast. *Geotimes,* v. 26, no. 8, August 1981, p. 22.)

"The progressive downward movement of unconsolidated materials into underground limestone openings naturally would affect the surface of the land and produce funnel-shape sinkholes at the land surface.

". . . Where the unconsolidated overburden is thick, say in the range of 30 to 500 ft., the likelihood of seeing funnel sinkholes developed at the surface in response to the downward migration of unconsolidated particles in limestones at great depth is much less likely. That is not to say that there is no downward migration of particles into underlying cavernous openings within the limestone. In fact, drilling into such limestones often establishes the presence of large amounts of sand, clay, and silt within cavernous openings.

". . . The downward movement of particles into cavernous openings frequently results in another cavity developing entirely within the unconsolidated materials above the bedrock, with its lower point connected to an opening into the limestone. Doubtless it begins as a small cavity. As material falls from the roof of this cavity it may migrate downward into the limestone, and hence the upper cavity enlarges. If the upper cavity, in unconsolidated materials, is within the zone of saturation, movement of material from the roof to the bottom of the cavity and on down into the limestone may be very slow. At the surface, one would not know that an open cavity existed within totally unconsolidated material *above* the bedrock. However, with the continued slow enlargement of the cavity due to particles occasionally falling from the cavity roof, it may grow to a size so large that the overlying lithostatic load (the weight of all the material between the top of the cavity and the land surface) can no longer be supported. At that moment the cavity collapses—a catastrophic event! . . ."

FIGURE 3.23 *Mammoth Cave Map*

Part of the U.S.G.S. Mammoth Cave quadrangle, Kentucky, 1922. Scale, 1:62,500; contour interval, 20 feet.

Name Section Date

EXERCISE 14C
EVOLUTION OF A KARST TERRAIN

The area covered by the Mammoth Cave map (fig. 3.23) is a good example of the effect of underlying strata on the topography. A cursory inspection of the map reveals that, from a topographic point of view, the southern one-third of the map area is markedly different from the northern two-thirds. The reason for this is readily apparent from the geologic cross-section in figure 3.24. This north-south cross-section passes through the main road intersection in the town of Cedar Spring near the center of the map, and its horizontal scale is the same as the scale of the map. The pronounced change in topography on the map is marked by the Dripping Spring Escarpment lying just north of and roughly parallel to the Louisville and Nashville Road.

1. Draw the line of the geologic cross-section on the map in black pencil.
2. Generally speaking, the area north of the Dripping Spring Escarpment and *west* of the line of the cross-section is characterized by an integrated stream system. Use a blue pencil to trace the drainage lines occupied by permanent and intermittent streams. Use the correct map symbol for each. Take care not to extend your blue lines beyond the limits of the streams as they are shown on the map. What is the dominant bedrock in the area drained by the stream system?

   ____

3. Why does the stream flowing north into Double Sink end so abruptly there?

   ____
   ____
   ____
   ____

4. Examine the topography *east* of the line of cross-section and north of Dripping Spring Escarpment. A number of valleys such as Cedar Spring Valley, Woolsey Hollow, and Owens Valley resemble stream-cut valleys, but there are no streams flowing in them. Account for the fact that the topography of this area resembles a stream-dissected terrain even though no streams occupy the existing valleys.

   ____
   ____
   ____
   ____
   ____

(Continued)

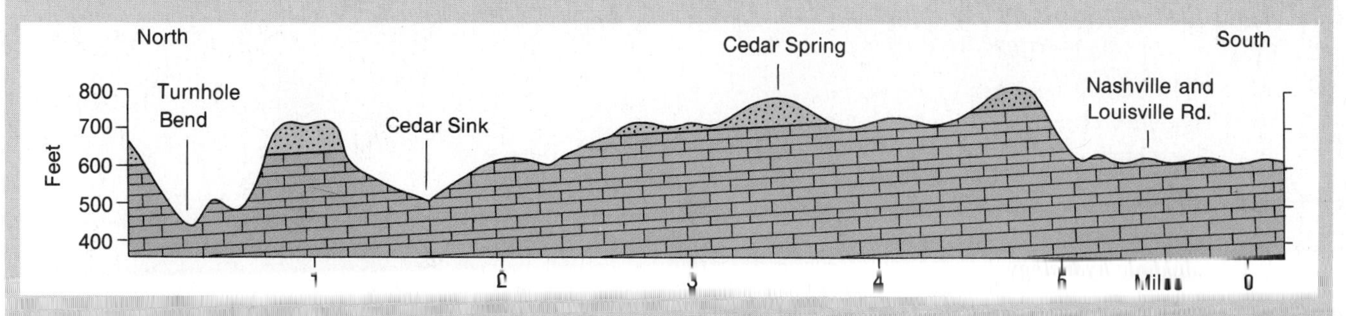

FIGURE 3.24

North-south geological cross section from Turnhole Bend on the Green River to a point about three-fourths of a mile south of the Louisville and Nashville Road, Mammoth Cave map. Vertical scale exaggerated about 10 times.
(Source: Simplified from The Geology of the Rhoda Quadrangle, Kentucky, USGS Map GQ-219 (1963).)

EXERCISE 14C CONTINUED

5. In making a comparison of the terrains east and west of the line of the cross-section (fig. 3.23), which of the following statements is most likely the correct one?
   (a) The area *west* of the line of cross-section will eventually resemble the area *east* of the cross-section as the streams cut deeper into the sandstone and encounter the underlying limestone.
   (b) The area *east* of the line of cross-section will eventually resemble the area *west* of the cross-section as the overlying sandstone is further eroded.

   ____
   ____
   ____
   ____
   ____

6. Assume that the sandstone formation in figure 3.24 had an original thickness of 100 feet. On figure 3.24, draw the upper and lower contacts of the sandstone formation as it existed at some time in the geologic past before stream erosion began stripping it away.

Web Connections

http://www.nps.gov/maca/karst.htm

Features of karst topography with definitions of terms.

http://www.epa.gov/ogwdw/index.html

Links to a variety of information sources related to drinking water quality, drinking water standards, etc.

http://www.dos.state.fl.us/dhr/bar/hist_contents/karst.html

Block diagrams showing a modern Florida Aquifer, a perched water system, and karst sinkhole hydrology.

http://geolab.muc.edu/GY111/MamCv.SU.html

A short exercise dealing with the Mammoth Cave, Kentucky, Quadrangle.

Glaciers and Glacial Geology

A *glacier* is a mass of flowing land ice derived from snowfall. The two major types of glaciers are alpine or valley glaciers and continental glaciers or ice sheets. An *alpine glacier* is one that is confined to a valley and is literally a river of ice. An *ice sheet* or *continental glacier* covers an area of continental proportions and is not confined to a single valley. Landforms produced by both alpine and continental glaciers are distinctive features that can be recognized on aerial photographs and topographic maps.

Glaciology is the study of snow and ice. Alpine or valley glaciers occur in mountain valleys where adequate snowfall sustains them. Most of this snow falls during the winter and covers the entire glacier. During the ensuing summer, some or all of the previous winter snowfall is melted and is discharged from the glacier in meltwater streams. The snow that remains at the end of the summer melt season gradually changes to ice and becomes part of the glacier.

Wastage and Accumulation

The 12-month period of winter snow accumulation (accumulation) and summer melting (wastage) is known as the *budget year of a glacier.* In either hemisphere, the budget year begins at the end of the melt season just before the first snows of the winter and ends about 12 months later at the end of the wastage season.

During the wastage months of the budget year, the previous winter's accumulation is partly removed from the upper reaches of the glacier. The snow that remains there lies in the *zone of accumulation.* Over the lower reaches of the glacier, the previous winter's snowfall is completely removed, and some of the underlying glacier ice is also melted. The area of the glacier that suffers wastage of both snow and ice is called the *wastage zone.* The line that separates the accumulation and wastage zones for a given budget year is the *annual snow line.* This is shown by a dashed line on the photograph of the Snow Glacier in figure 3.25 that was taken at the end of the wastage season. The white area above the snow line of the Snow Glacier is the residue of the previous winter's snowfall that survived the summer melting. On the lower reaches of the Snow Glacier, all of the winter snow was melted so that bare glacier ice is exposed. Some of this ice has also been lost by summer melting. The streams flowing from the glacier terminus in the lower right-hand corner of the photograph are fed by melted snow and ice from the glacier.

Glacier Mass Balance

It is possible to measure both wastage and accumulation for a given glacier during a budget year. The relationship between the annual wastage and accumulation over a period of years describes the general health of a glacier. For example, if wastage exceeds accumulation for a period of 10 to 20 years, the total mass of the glacier will decrease, a condition usually reflected in the retreat of the glacier terminus and a reduction in thickness of the glacier. If accumulation exceeds wastage for several years, the glacier terminus advances and the glacier thickens.

A comparison between accumulation and wastage for a given budget year yields the *net mass balance* of the glacier. Net mass balance is expressed numerically in feet or meters of water equivalent and represents a hypothetical layer of water determined by many measured thicknesses of columns of snow or ice on the glacier's surface. If a glacier gains in mass during a budget year (i.e., accumulation exceeds wastage), the glacier is said to have a *positive net mass balance.* If the glacier loses mass during the budget year, it has a *negative net mass balance.*

Alpine Glaciers

An alpine glacier erodes the floor and walls of the valley through which it flows. Furthermore, it functions as a conveyor belt that transports debris from the valley floor and walls to the glacier terminus where it is deposited. Debris carried and deposited directly by a glacier is *till,* an unsorted mixture of particles ranging in size from clay to boulders.

Till accumulates in various topographic forms called *moraines.* Debris eroded from the walls of an alpine glacier is transported along the margin of the glacier as a *lateral moraine.* The joining of two lateral moraines at the confluence of a tributary glacier and the main glacier forms a *medial moraine* (fig. 3.26). Both lateral and medial moraines ultimately reach the glacier terminus or snout to form an *end moraine.*

Moraines are identified on photos of glaciers as dark bands in the wastage zone (fig. 3.25). Moraines are not visible in the accumulation zone because they are covered by the perennial snow that exists there. Lateral and medial moraines define the flow lines of an alpine glacier and cannot cross each other, but they may converge toward each other near the glacier terminus. An end moraine can remain long after the terminus where it was formed has retreated. Successive end moraines lying beyond the snout of a retreating glacier are called *recessional moraines.*

Glaciers that terminate in a lake or the ocean produce *icebergs,* masses of glacier ice that become detached from the glacier terminus, a process called *calving.* Icebergs float freely but can become grounded when they are carried to shallow water by wind and currents. Icebergs eventually melt.

FIGURE 3.25

Oblique aerial photo of Snow Glacier, Kenai Peninsula, Alaska. Annual snow line shown as dashed line.

(Photograph by Austin Post, U.S.G.S.)

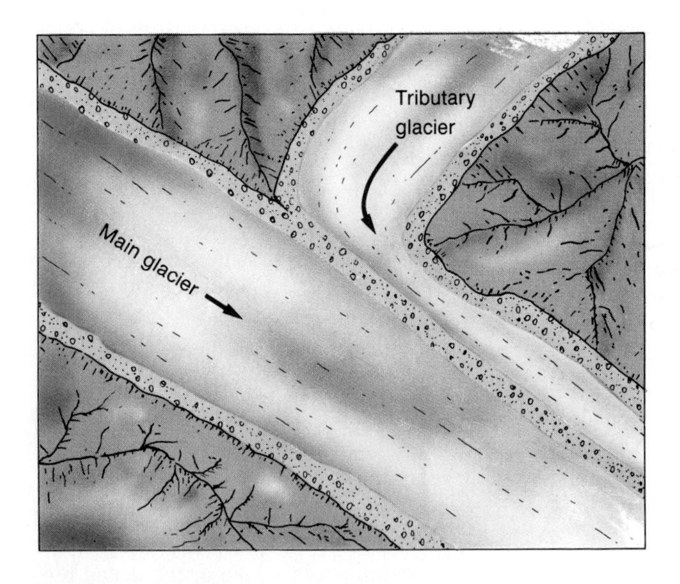

FIGURE 3.26

Lateral moraines join to become a medial moraine as a tributary glacier joins the main ice mass.

(From Carla W. Montgomery, Physical Geology, 3d ed. Copyright © 1993 McGraw-Hill Company, Inc., Dubuque, Iowa. All rights reserved. Reprinted by permission.)

LANDFORMS PRODUCED BY ALPINE GLACIERS

Erosional features usually dominate a terrain that was shaped by valley glaciers. A glacially eroded valley is *U-shaped* in cross-section, and its headward part may contain a *cirque,* an amphitheaterlike feature. A cirque that contains a lake is called a *tarn. Hanging valleys* form where tributary glaciers once joined the trunk glacier, and waterfalls cascade from the hanging valleys to the main valley floor.

A narrow, rugged divide between two parallel glacial valleys is an *arête,* and the divide between the headward regions of oppositely sloping glacial valleys is a *serrate divide.* A pyramid-shaped mountain peak near the heads of valley glaciers or glaciated valleys is a *horn,* the most famous of which is the Matterhorn near Zermatt, Switzerland.

Moraines left by former alpine glaciers are rarely visible on standard topographic maps because the contour interval is usually larger than the height of most moraines.

WEB CONNECTIONS

http://ak.water.usgs.gov/glaciology/

The USGS makes measurements re mass balance for not only South Cascade Glacier but also Gulkana and Wolverine Glaciers in Alaska. Annual and seasonal runoff is also available along with daily air temperature and precipitation measurements.

http://www-nsidc.colorado/

Home page for the National Snow and Ice Data Center. Click on Education Resources for specific information on glaciers, snow, avalanches, and other snow and ice information including facts and Q&A.

http://www.nsf.gov/home/polar/start.htm

Home page for polar research sponsored by the National Science Foundation. Index provides access to a broad array of sites dealing with science including geology and glaciology at both poles.

EXERCISE 16C

ICE-CONTACT DEPOSITS

Figure 3.37 is a stereopair of an esker in Michigan.

1. Trace the crest of the esker in red pencil.
2. Visualize the esker crest in profile. Use a black pencil to draw a single closed contour line showing the highest part of the esker crest. Draw the contour line on the right-hand photo while viewing the stereopair in stereovision.
3. Just east of the bend in the esker is a conical hill or knob. What is the name for this ice-contact deposit?

   ___
4. What constructional materials might be available from the esker and the ice-contact deposit?

   ___

   ___

Figure 3.38 shows two north-south trending eskers.

5. Trace the crest of each esker with a light-colored, felt-tip highlighting pen.
6. What local name is used for the eskers?

   ___

   ___
7. What map evidence verifies the assumption that eskers are ice-contact deposits?

   ___

   ___
8. The crest of the esker east of the Penobscot River does not lie at a constant elevation along its course. Given this observation, why is it necessary to invoke the presence of glacier ice to account for the origin of an esker?

   ___

   ___

   ___

   ___

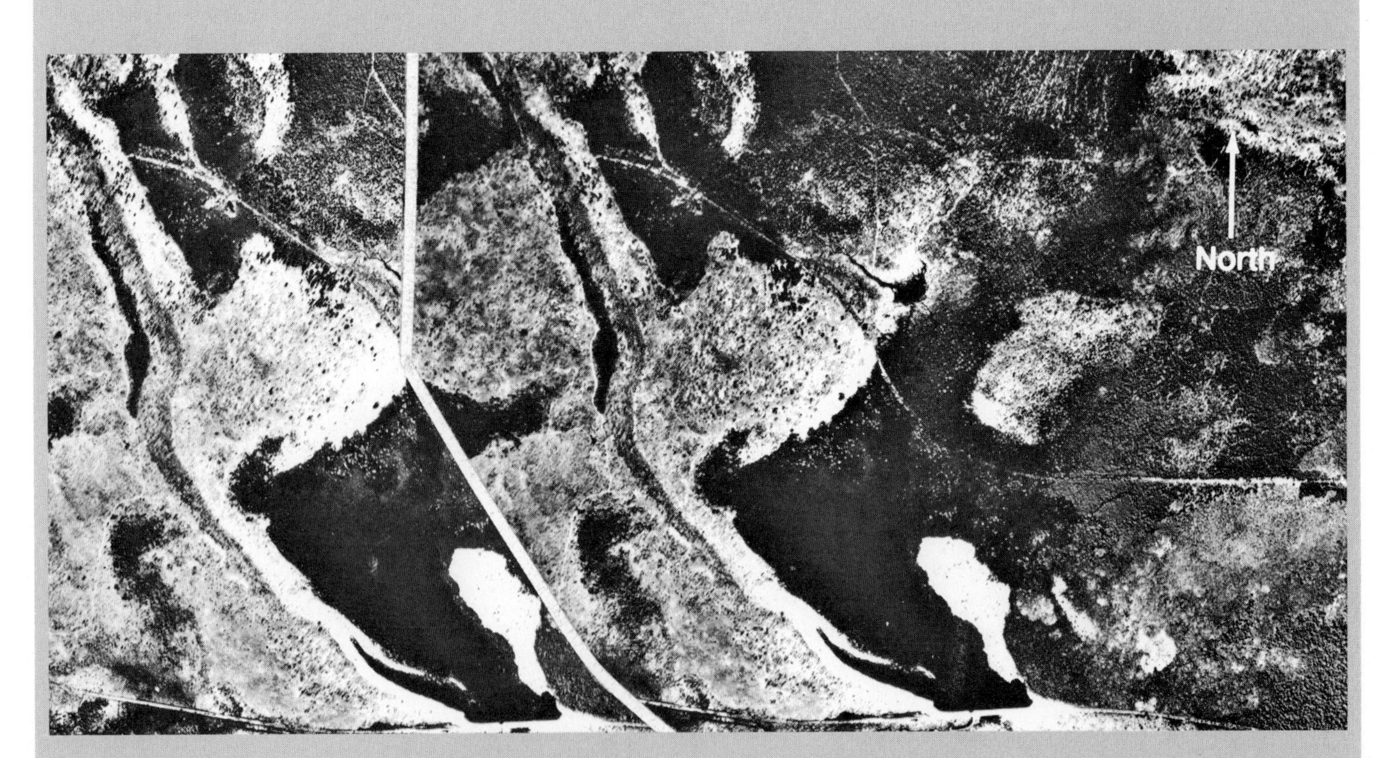

FIGURE 3.37

Stereopair of part of an esker in Michigan. Scale, 1:24,000.
(Courtesy of U.S. Geological Survey.)

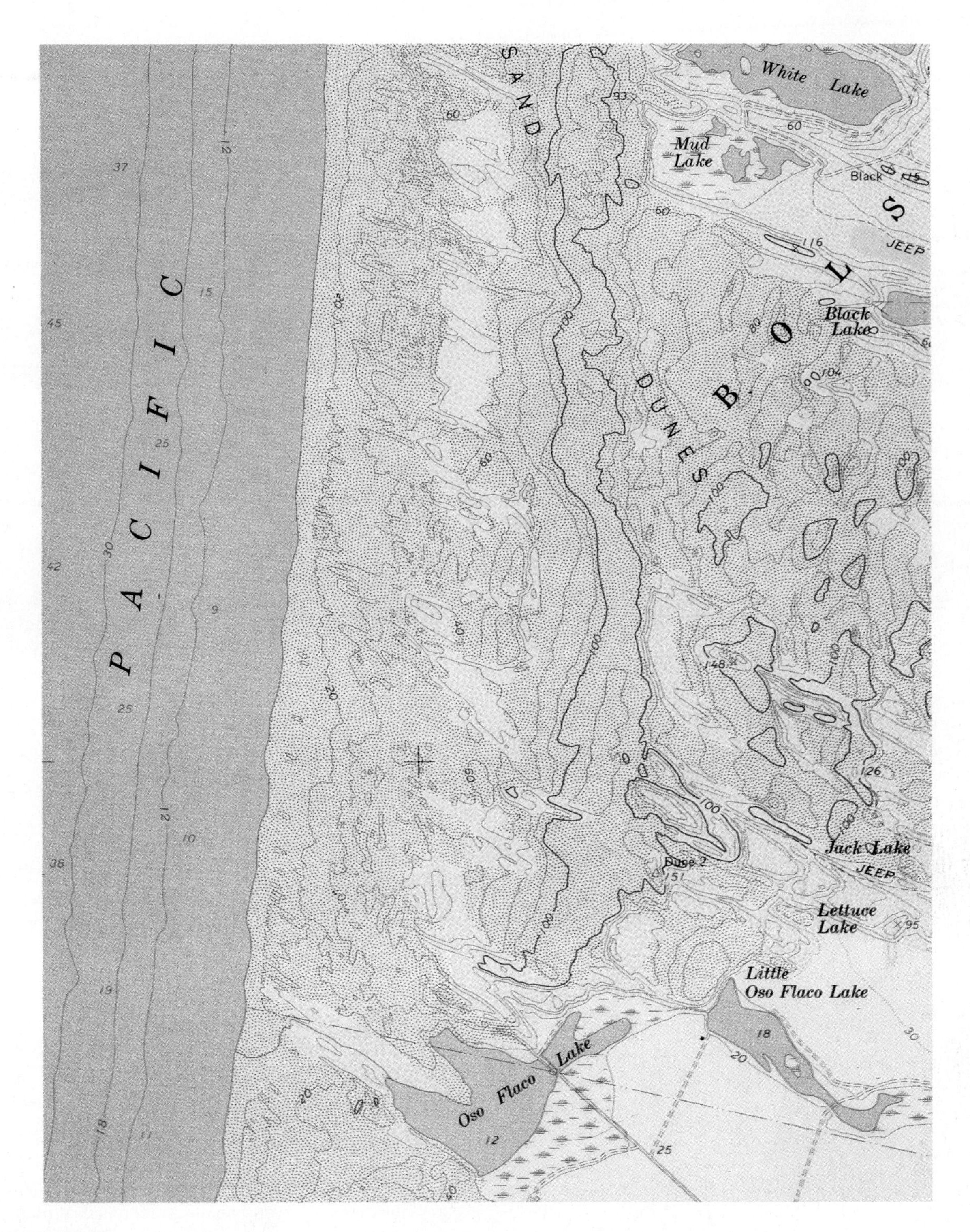

▲ **FIGURE 3.43** *Oceano Map*

Part of the Oceano quadrangle, California, 1965. Scale, 1:24,000; contour interval, 20 feet.

◀ **FIGURE 3.42** Stereopair of sand dunes on the California Coast north of Santa Barbara. Scale, 1:28,800. (Photos taken June 27, 1963. Series GS-VASJ, numbers 1–262 and 1–265.) This stereopair will be used again in Exercise 18A. Please keep it to complete that exercise.

EXERCISE 17B CONTINUED

4. What is the source of the sand contained in the dunes?

5. What is the likely origin of Little Oso Flaco Lake and Mud Lake?

EXERCISE 17C
INACTIVE DUNE FIELDS

Figure 3.44 is a Landsat image, similar to a black-and-white aerial photograph, taken from an altitude of 570 miles over western Nebraska in the winter of 1973. This area is known as the Sand Hills because the topography consists of a massive dune field that is now virtually inactive because of a lush vegetational cover of prairie grasses. The dunes were formed during a time when the rainfall was less than at present and unable to support the grasses that grow there now. A heavy cover of snow enhances the topography.

The relief of the Landsat may appear inverted. If the dune ridges appear to be depressions, turn the photo upside down so that the north arrow points toward you and the true relief will "pop" into view.

The rectangular area in the northwestern corner of the map is the area covered by the Ashby map of figure 3.16. The two dark bands in the lower part of the image are the North and South Platte Rivers which converge toward the town of North Platte, Nebraska, near the eastern margin of the image.

1. What is the R.F. of the Landsat image?

2. What are the only features visible on both the Landsat image and the Ashby map?

3. Some of the interdune depressions visible on the Ashby map contain lakes. Assuming that some of the interdune depressions were sites of deflation during the period of dune formation, what does the presence of the lakes imply about the elevation of the water table when the dunes were active?

4. The topography of the dunes on the Ashby map reveals that many of them are irregular in map view and contain no clear-cut distinction between gentle and steep slopes. However, those that are elongated in a general east-west direction are typical of those visible on the Landsat image. Study the contour lines on the north and south side of the east-west trending dunes on the Ashby map, determine which slopes are the steepest, and infer the direction of the prevailing winds at the time of deposition. ______

5. Based on the answer to question 4, draw arrows about ¼ inch long at several points on the Landsat image to show the direction of the prevailing winds when the dune field was formed.

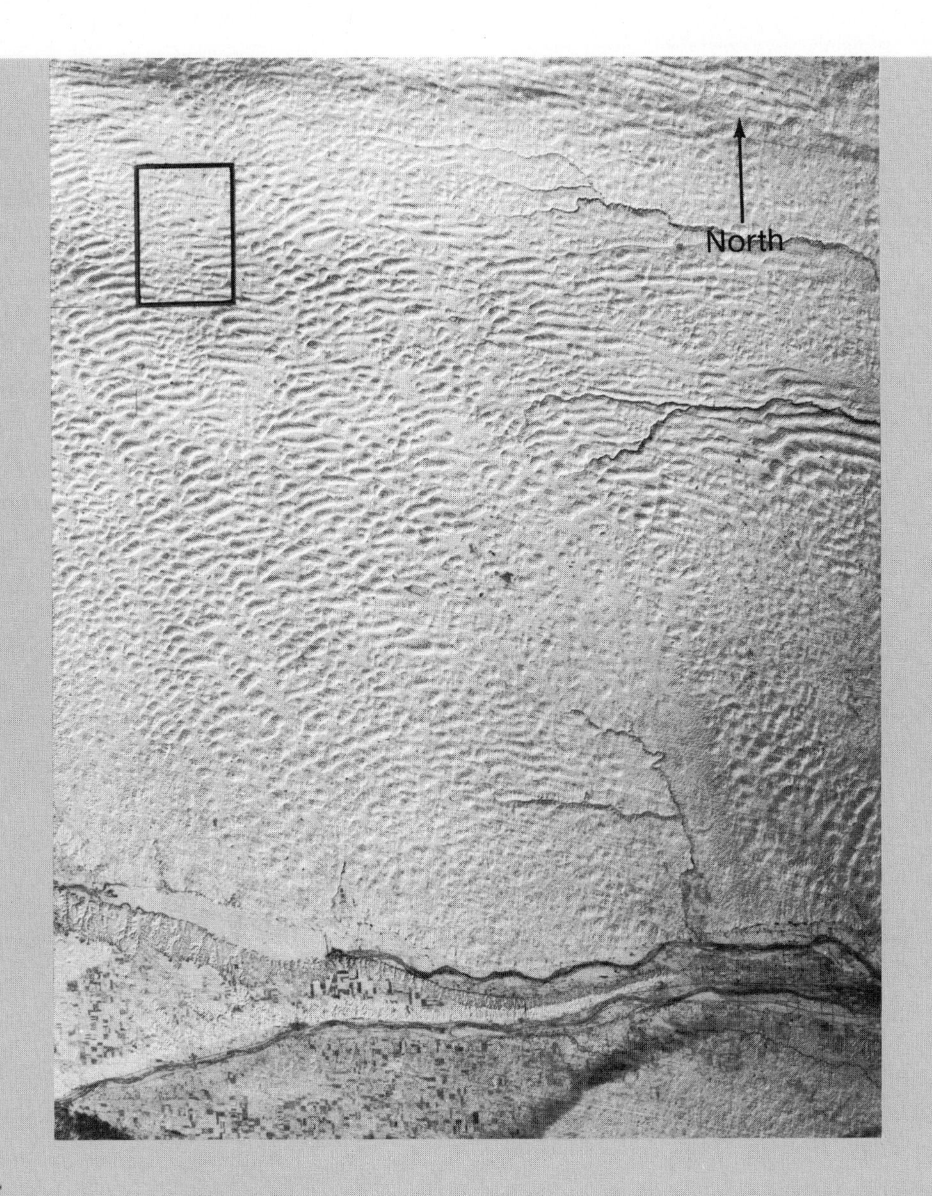

FIGURE 3.44

Landsat image of part of western Nebraska made on January 9, 1973 from ERTS-1 at an altitude of 570 miles. The rectangle in the northwest corner is the area covered by the Ashby map of figure 3.16.

(NASA ERTS E-1170-17020, Courtesy of NASA and the U.S.G.S. EROS Data Center, Sioux Falls, South Dakota 57198.)

Web Connections

http://www.nps.gov/grsa/

National Park Service page for Great Sand Dunes National Monument, Colorado. Click on Natural and Cultural History for discussion of the geology.

http://www.nps.goc/whsa/ndxtxt.htm

White Sands National monument is a unique dune area with the sand size particles made up of gypsum. Information about the area can be found at this site.

http://pubs.usgs.gov/gip/deserts/index.html

One of the excellent USGS public information booklets dealing with deserts, dunes, etc.

http://geology.wr.usgs.gov/MojaveEco/dustweb/dusthome.html

Dust studies in Southern Nevada and California. The importance of dust in soils and geology is discussed. Other related Web Sites are provided.

http://ceres.ca.gov/ceres/calweb/coastal/dunes.html

Brief discussion of coastal dunes in California. Guide to other sources of information on this topic.

Modern and Ancient Shorelines

The interaction between the ocean or a large lake and the land occurs at the coastline or shoreline. Waves generated by winds eventually reach the shore, where the wave energy is dissipated. Waves breaking along a shore are geologic agents of erosion, transportation, and deposition. This section deals with the modern and ancient landforms produced by one or a combination of these agents as they are portrayed on topographic maps and aerial photographs.

Depositional Landforms Produced by Wave Action

A wave crest moving toward shore is bent or *refracted* as it approaches shallow water (fig. 3.45). As the successive incoming waves break just off shore, a *longshore current* is initiated (fig. 3.46). A wave breaking on the shore carries sand particles up the beach in the direction of wave movement. When the water from the spent wave flows back down the beach, it follows a course controlled by the slope of the beach. Sand particles exposed to this alternate movement are moved along the beach, a process known as *beach drift* (fig. 3.46). The longshore current moves fine sand and beach drift moves coarser sand particles parallel to the shoreline. When the longshore current slows due to deeper water associated with an indentation on the coast, such as a bay or estuary, sand is deposited in the form of a *spit.* A spit may grow across the mouth of a bay to form a *baymouth bar* (fig. 3.47). A spit that is curved shoreward is a *recurved spit.* Other sandy features along the coastline include *beaches* and *barrier islands.* Indentations along the coastline that become isolated from the main body of water become lakes or lagoons. *Lagoons* are shallow water bodies lying between the main shoreline and a barrier island. A *barrier island* is an elongate sand island parallel to the shoreline. A break or passageway through a bar or barrier island is a *tidal inlet,* so called because it allows currents to flow into the lagoon during rising tides and out of the lagoon during falling tides. Lagoons and coastal lakes eventually become filled with sediment and are transformed into *mud flats, marshes,* and *wetlands.*

Beach sand is attacked by onshore winds that carry the sand inland to form coastal dunes. The sand lost to wind action is replenished by wave action and beach drift.

Erosional Landforms Produced by Wave Action

An initial shoreline consists of bays and headlands. A *headland* is a part of the coast that juts out into the lake or ocean. Wave refraction concentrates the energy of waves against headlands (fig. 3.45). The landforms produced by this intensified wave action are *wave-cut cliffs,* seaward-

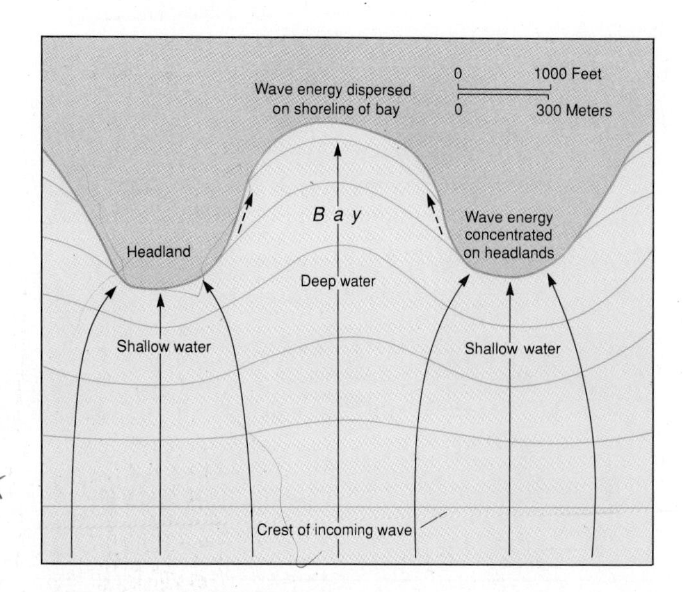

FIGURE 3.45

Schematic map showing the refraction of waves approaching an irregular shoreline. Wave energy is concentrated on the headlands and dispersed in the bays. The arrows represent lines of equal wave energy. These are equally spaced where the water depth is below the wave base but curved toward the headlands when one part of the wave crest strikes shallow water before the rest of the wave. Dashed arrows show the direction of beach drift.

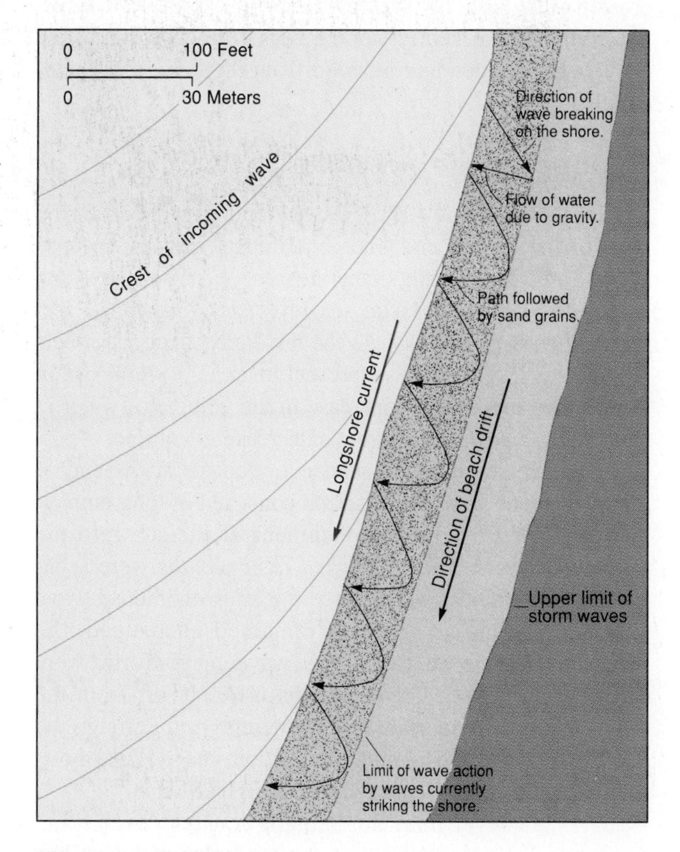

FIGURE 3.46

Schematic map showing the relationship of refracted waves breaking on the shore to the longshore current and beach drift.

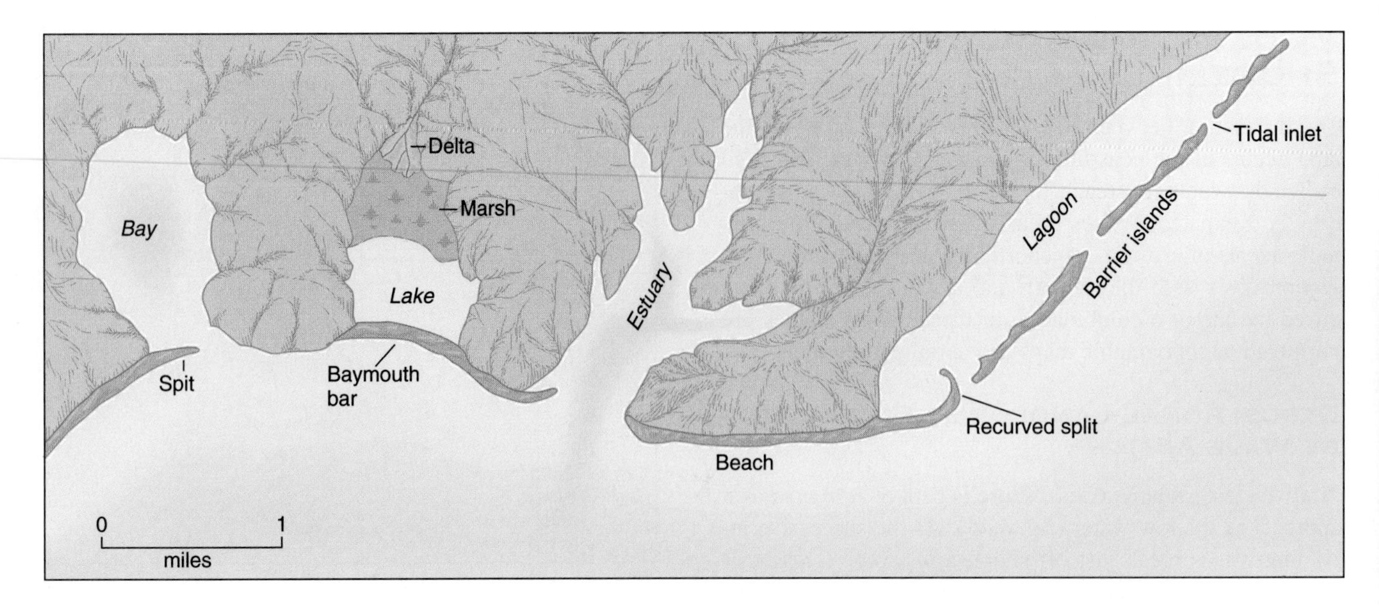

FIGURE 3.47

Schematic map showing landforms associated with wave action and longshore drift of sand (from left to right in the figure).

facing escarpments formed by wave erosion at their bases, and *stacks,* rocky pillars that are remnants of retreating wave-cut cliffs.

As a wave-cut cliff retreats under constant wave erosion at its base, a wave-cut platform is formed. A *wave-cut platform* is a gently sloping rock surface lying below sea level and extending seaward from the base of a wave-cut cliff.

Evolution of a Shoreline

Sea level was about 400 feet lower than at present during parts of the Pleistocene when continental glaciers covered about 30% of the earth's land surface. We also know that during the last interglacial sea level stood about 20 feet higher than it does today. As the last Pleistocene ice sheets melted, sea level rose to its present level. A *eustatic rise* in sea level results from an increase in the volume of water in the oceans, a *eustatic fall* from a decrease in volume.

The shorelines around many coasts at the end of the Pleistocene were irregular and consisted of long embayments formed by the encroachment of the sea into the mouths of rivers. These *drowned river mouths* were separated by headlands, which were the sites of intense wave erosion. So long as sea level remained unchanged, the headlands became wave-cut cliffs, wave-cut platforms were formed, and drowned river mouths were cut off from the sea by the growth of spits and baymouth bars. Through the process of wave erosion and deposition, the original post-Pleistocene shoreline changed to a straighter form.

Sea level does not remain constant over time, however. A local uplifting of the land along a coast has the same effect as a drop in sea level. Because it cannot always be determined whether sea level has risen or fallen eustatically, whether the land has risen or subsided, or whether a combination of these events has occurred, it is common practice to refer to *relative* changes in sea level.

Generally, a relative fall in sea level produces an *emergent shoreline,* and a relative rise in sea level produces a *submergent shoreline.* The shorelines of the world's coasts at the end of the Pleistocene were generally submergent, but in coastal regions that were formerly near or within the borders of the continental glaciers, the land began to rise in response to the retreat of the ice caps whose weight had depressed the earth's crust. This uplift did not begin immediately upon retreat of the ice, so there was time for erosional and depositional shore features to be built. These shoreline features now lie above modern sea level, where they are exposed to the normal processes of subaerial weathering and erosion which modify and eventually destroy them.

Deltas

A *delta* is a nearly flat plain of riverborne sediment extending from the river mouth to some distance seaward. The deltaic sediments are deposited by *distributaries,* channels that branch out from the main channel of the river. Distributaries are bordered by *natural levees,* narrow ridges of river sediment built by a river overflowing its banks during flood stages. A delta with many distributaries is called a *birdfoot delta* because of its resemblance in plan view to the outstretched claws of a bird's foot.

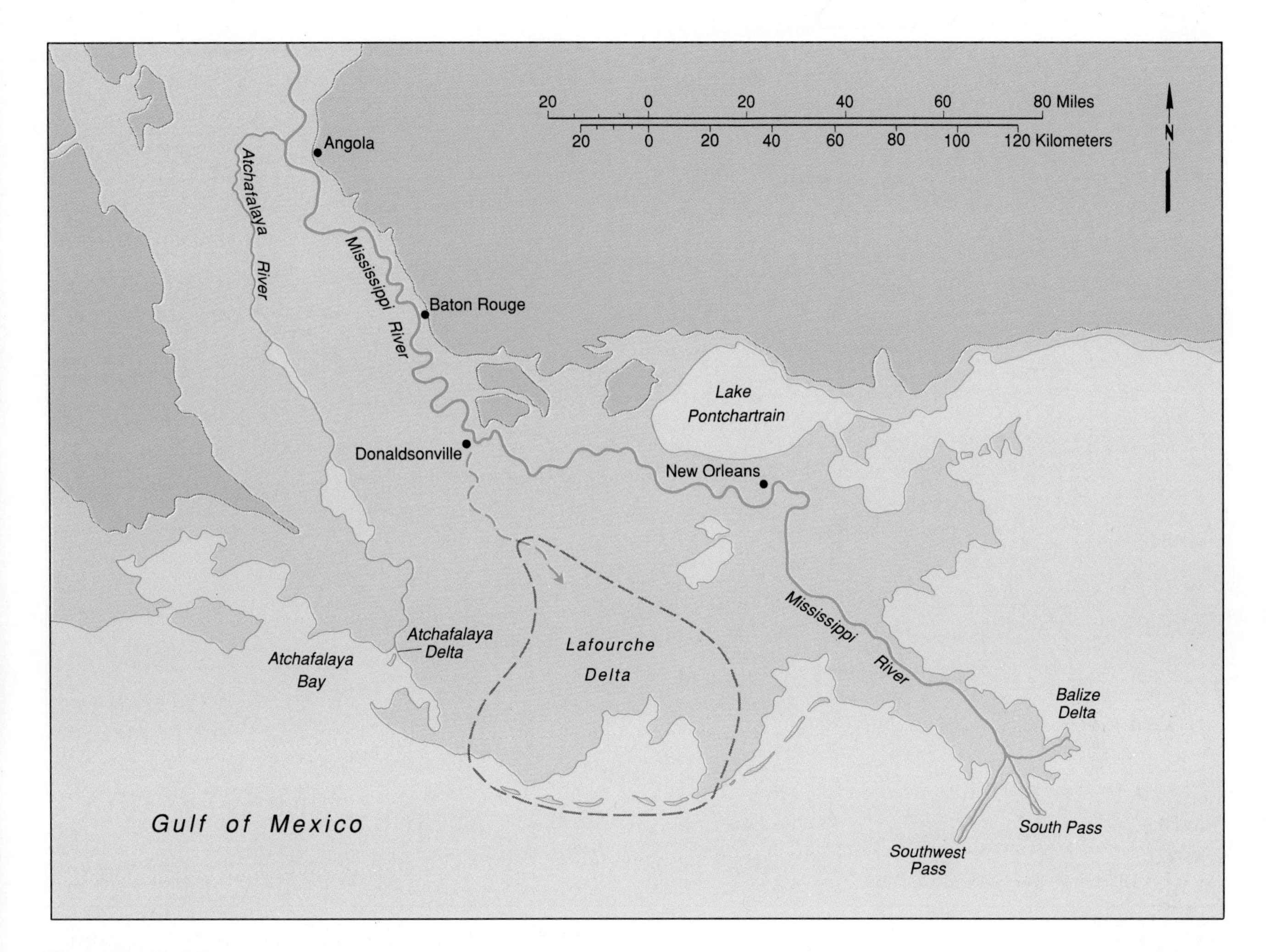

FIGURE 3.48

Map of the Mississippi River valley and deltas associated with it.

(After Regional Geomorphology of the United States by W. D. Thornbury. Copyright © 1956 John Wiley & Sons.)

New distributaries are formed from time to time, and older distributaries are abandoned. An entire delta may be abandoned when the course of the river feeding it shifts to a new course many miles upstream. The abandoned delta then becomes an *inactive delta.* The inactive delta no longer receives sediment and is modified by wave action to the extent that its seaward margin becomes smooth and it loses its characteristic birdfoot form.

An active delta expands seaward if the supply of sediment is too large to be dissipated by wave action and longshore currents. A shoreline that moves seaward by deltaic growth is called a *prograding shoreline.* Progradation ceases when a delta becomes inactive, and its seaward margin may in fact retreat due to wave action and subsidence of the deltaic deposits. Eventually, waves and currents redistribute sand along the old delta front to create lagoons behind barrier islands, tidal inlets, and saltwater marshes. An inactive delta can be reactivated if the river course shifts upstream and discharges once again into the site of the former active delta.

The Mississippi Delta

The Mississippi River drainage system carries an enormous load of sediment that eventually is dumped into the Gulf of Mexico along the coast of Louisiana. The delta currently being built by the Mississippi River is the Balize or Birdfoot delta, and it has been functional during the last 800 to 1,000 years. Over the last 7,500 years, four previous deltas, now inactive, have been built by ancestral courses of the Mississippi River. The most recent of these inactive deltas, the Lafourche delta, was active between 2,500 and 800 years ago. It lies adjacent to and on the west side of the Balize delta (fig. 3.48).

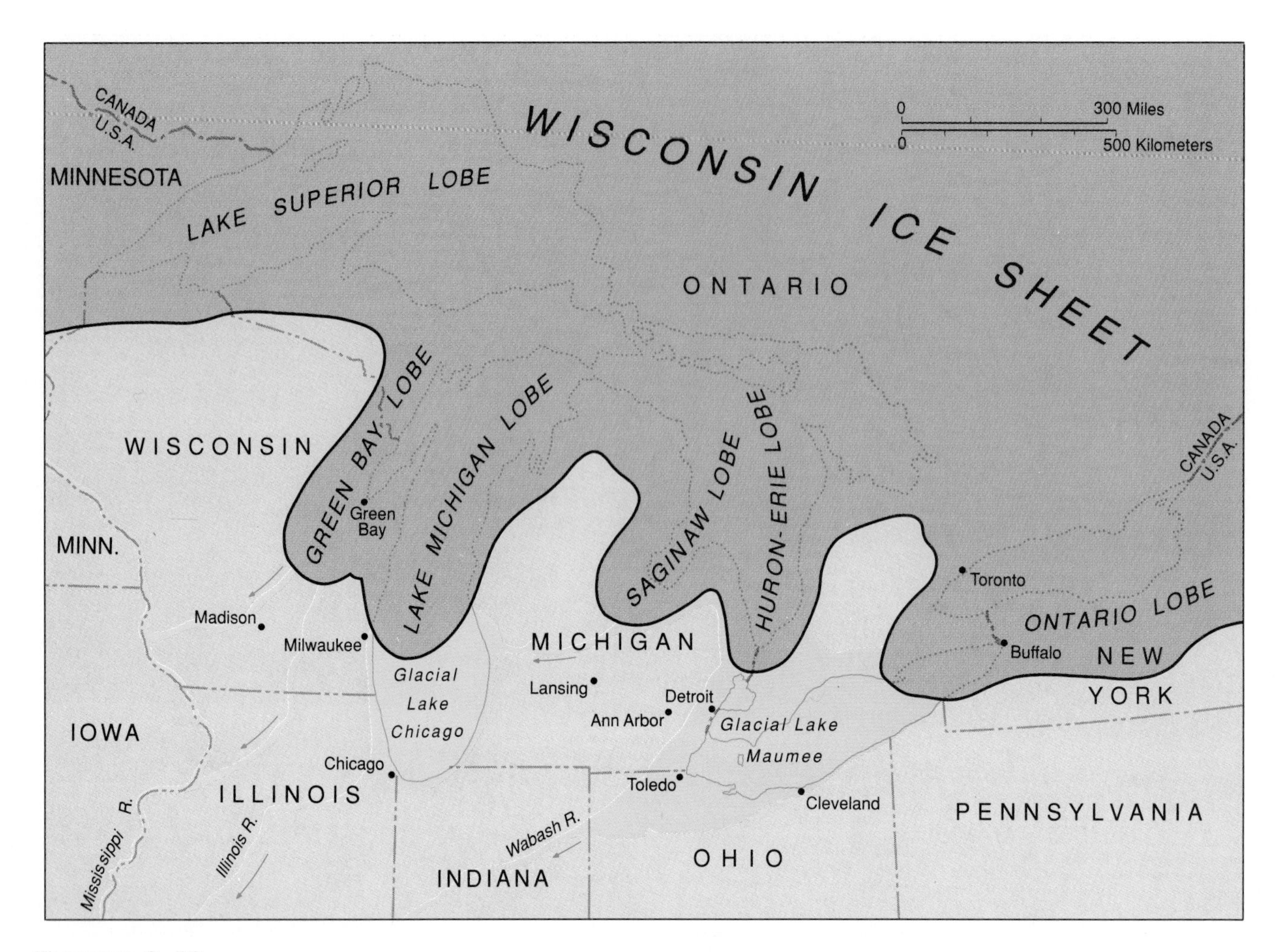

FIGURE 3.49

Map showing the extent of the last, or Wisconsin, ice sheet in the Great Lakes region about 14,000 years ago. The glacial lakes Chicago and Maumee drained through outlets to the Gulf of Mexico via the Mississippi River.

(After Elements of Geology, 3d ed., by J. H. Zumberge and C. A. Nelson, Copyright © 1972 John Wiley & Sons Inc.)

In the mid-twentieth century, the Mississippi River was discharging about 25 percent of its flow through the Atchafalaya River, some 100 miles north of Atchafalaya Bay (fig. 3.48). This flow would have increased until the entire volume of the Mississippi River would have discharged into the Gulf of Mexico through the Atchafalaya River. Had this natural event occurred, the modern course of the river south of Angola and the Balize delta would have been abandoned. However, this natural diversion was forestalled in 1963 when the U.S. Army Corps of Engineers completed a control dam at the diversion site. In spite of this human intervention, however, the sediment deposited by the Atchafalaya River into Atchafalaya Bay has started what might become the next delta of the Mississippi River.

Ancient Shorelines around the Great Lakes

During the retreat of the continental glaciers from the Great Lakes region, precursors of the modern Great Lakes formed around the lobate front of the ice margin (fig. 3.49). The shorelines of these ancestral Great Lakes are preserved in beach ridges and wave-cut cliffs that now lie above the elevations of the present Great Lakes. Several different ancestral lakes were formed at various times during the retreat of the ice, and each of these lake stages has been assigned a name to distinguish one from the other. Figure 3.49 shows the extent of glacial lakes Maumee and Chicago in the basins now occupied by Lake Erie and Lake Michigan. The levels of these and other stages can be determined by using the elevations of their shorelines to infer the elevations of the water planes of the lakes. For example, the base of an abandoned wave-cut cliff would mark the water plane of the lake that produced it.

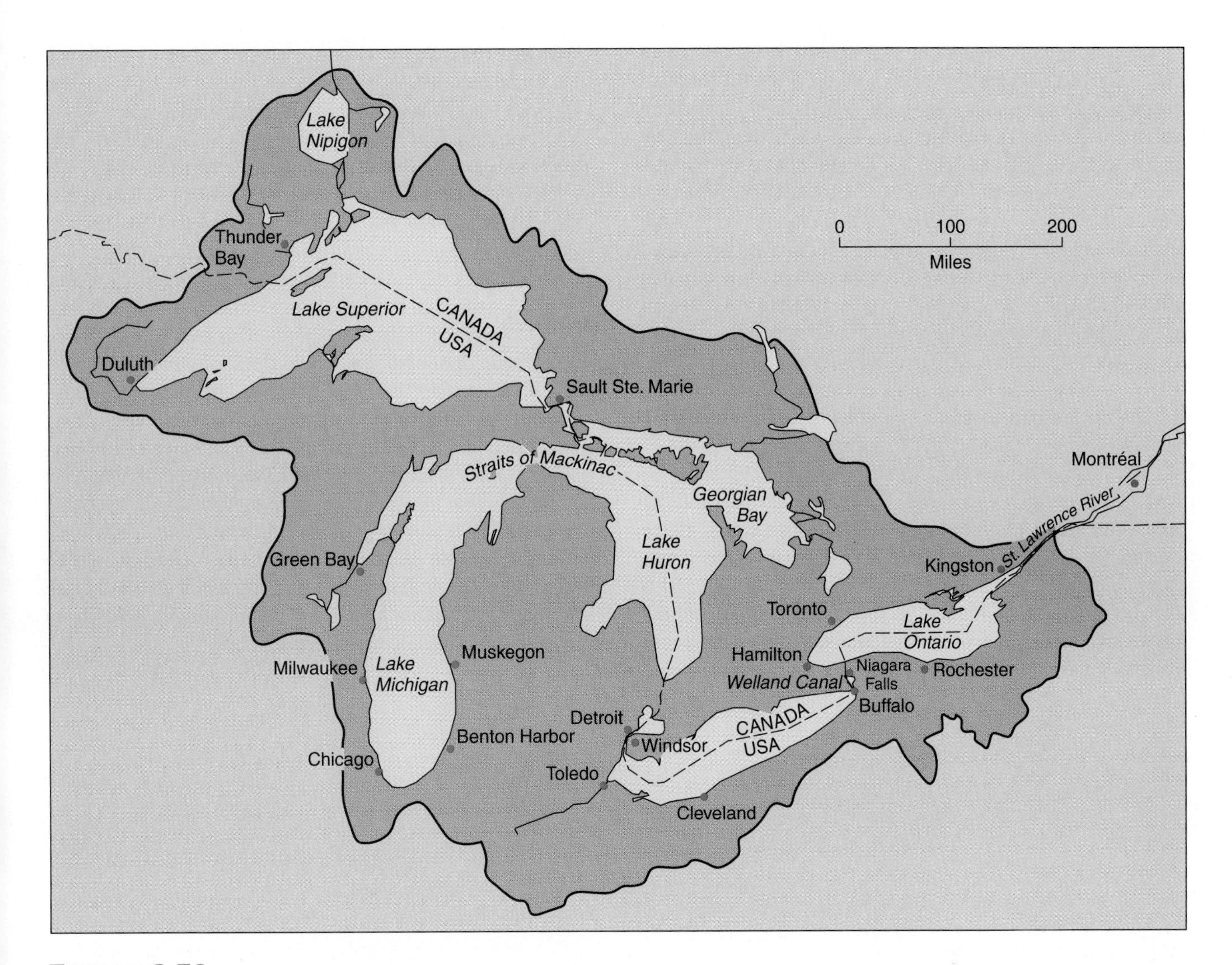

FIGURE 3.50

Map of the Great Lakes and the land area supplying runoff to them. The five lakes have a combined surface area of almost 100,000 square miles, and the drainage basin containing them is roughly twice that size.

Where two or more ancestral shorelines are related to the basin of a modern lake, we will assume that the highest shoreline is the oldest and the lowest is the youngest. While this relationship is not universally true for all the former lake stages, it will suit our purposes for this discussion. Note that a higher lake may have a different outlet; for instance, Lake Maumee and Lake Chicago and the other drainage systems presented on figure 3.49 had outlets to the Mississippi River at this time, because the present outlet of the lakes through the St. Lawrence River was blocked by the Ontario lobe.

The Modern Great Lakes

The Great Lakes straddle the U.S.-Canadian border and have a combined surface area of almost 100,000 square miles (fig. 3.50). The lakes came into being during the retreat of the Pleistocene ice sheet that once covered the entire area more than 10,000 years ago. The volume of water in the modern lakes is mainly a function of surface runoff from the surrounding drainage basin, evaporation from the lakes themselves, and the outflow of water to the Atlantic Ocean via the St. Lawrence River. The volume of water stored in the Great Lakes is reflected in their water levels. These levels are published monthly for each lake by the U.S. Army Corps of Engineers (fig. 3.56) and are a matter of public record.

The shorelines of the Great Lakes are sites of thousands of vacation homes and permanent dwellings. Those with lake frontage are desirable from an aesthetic point of view, but they are in considerable jeopardy during periods of high water levels when storm waves destroy sandy beaches formed during low water stages and cause severe recession of the bluffs and cliffs by wave erosion at their bases.

Much of the eastern shoreline of Lake Michigan, for example, is characterized by steep bluffs that are formed in old sand dunes and other unindurated Pleistocene sediments. Some of the bluffs rise to more than 100 feet above the lake. Private dwellings on the tops of these bluffs were built during periods of low water when the shores between the bases of the cliffs and the water's edge were characterized by wide sandy beaches. Those who purchased or constructed homes during low water stages believed that the wide beaches fronting their properties were a permanent part of the landscape and provided adequate protection from any future wave erosion. These were false assumptions, as many who owned property on the shores of Lake Michigan learned at great cost during the period 1950 through 1987.

Three times during this period, 1952–1953, 1972–1976, and 1985–1986, the water levels of Lake Michigan stood at extraordinarily high levels, and twice during the same period, 1958–1959 and 1964, the levels were extremely low (fig. 3.56). The water level of 581.6 feet in 1986 was the highest on record since 1900, and the water level of 575.7 feet in 1964 was the lowest on record for the same period. Thus, in the 22-year period between 1964 and 1986, the level of the lake varied by about 6 feet, mainly by natural causes.

The damage to property during high water stages is all too apparent in the photograph of figure 3.57. The house in figure 3.57 was abandoned by the time it was photographed in 1986.

One might ask why these houses and hundreds of others like them were built in the first place. The answer lies in the lack of understanding of the relatively short time it takes for the lake level to change drastically and in the inability of anyone to predict future levels over a time period of a decade or so. No one would think of building a house next to the one in figure 3.57 today. The consequences of such folly are all too apparent. But, when the houses along Lake Michigan and other Great Lake shores were built during low water stages, they seemed secure from the damage and destruction to which they were subjected in later years.

The lesson to be learned from this is that those who contemplate purchasing or building homes should be aware of geologic hazards and should endeavor to acquire as much information as possible about building sites from the public record before proceeding.

FIGURE 3.51 *Point Reyes Map*

Part of the U.S.G.S. Point Reyes quadrangle, California, 1954. Scale, 1:62,500; contour interval, 80 feet.

Name Section Date

EXERCISE 18B
DELTAS OF THE MISSISSIPPI RIVER

Figure 3.52 shows the Balize delta of the Mississippi and the plume of sediment being discharged to the Gulf of Mexico. The major distributaries of the Balize delta are called *passes,* and their names and the names of associated interdistributary bays are shown in figure 3.53. The inactive Lafourche delta lies on the western flank of the Balize delta (see also fig. 3.48).

1. What are the narrow ridges visible on either side of Southwest Pass in figure 3.52?

____

____

2. The light blue color around the Balize delta in figure 3.52 is a plume of suspended sediment. What is the immediate resting place of this sediment?

____

____

3. If the Balize delta should be abandoned in favor of the Atchafalaya delta, describe the changes that would occur on the seaward edge of the Balize delta.

____

____

____

____

____

4. The Mississippi River is a major artery of commerce connecting the Gulf of Mexico with New Orleans and Baton Rouge. Describe the impact on the river between the delta and the cities along its course if human intervention had not been imposed on the site of natural diversion of the Mississippi River near Angola.

____

____

____

____

____

5. What features of the Lafourche delta indicate that it was once an active delta?

____

____

____

____

____

FIGURE 3.52

False color image of the Mississippi Delta made from Landsat on April 3, 1976.

(Courtesy of NASA and the U.S.G.S. EROS Data Center, Sioux Falls, South Dakota 57198.)

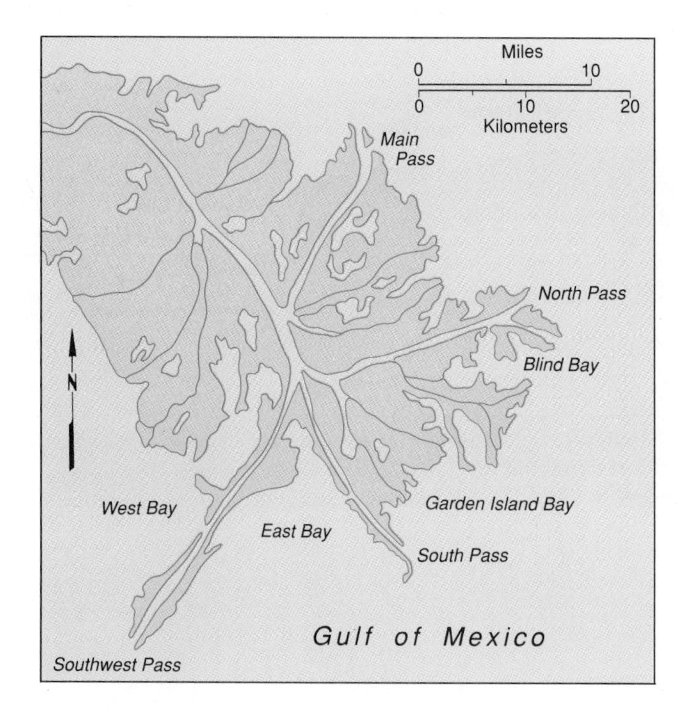

FIGURE 3.53

Map of the Balize or Birdfoot delta of the Mississippi River.

(From Orin H. Pilkey, et al., Coastal Land Loss Short Course in Geology, Vol. 2. Copyright © 1989 by the American Geophysical Union, Washington, DC.)

Name Section Date

EXERCISE 18C

ANCESTRAL LAKES OF LAKE ERIE

The North Olmstead map (fig. 3.55) covers an area just west of Cleveland, Ohio. Detroit Road and Center Ridge Road lie along two shorelines ancestral to Lake Erie. These will be referred to as the Detroit shoreline and the Center Ridge shoreline. The Center Ridge shoreline was formed about 13,000 years ago during the waning phases of the Pleistocene, approximately 1,000 years after glacial lake Maumee.

Figure 3.54 is a north-south profile showing the location of Center Ridge Road and Detroit Road. The profile is aligned along Forest View Road, which is just west of Forest View School in the upper right hand corner of the map. The generalized profile extends north into Lake Erie and south to the 740-ft. contour line.

1. Draw the line of profile with a black pencil on figure 3.55.
2. With reference to the profile, compare the Detroit and Center Ridge shorelines with the shoreline of Lake Erie. Which of these three shorelines are erosional and which are depositional?

____

____

____

3. Use a sharp pencil and a straightedge to draw the water surfaces of the lakes that produced the two ancestral shorelines. Write the elevations of each above the two lines you have drawn.
4. Draw a solid black line on the profile between Detroit Road and Wolf Road to show what the original profile might have looked like before the Detroit shoreline was formed.
5. Based on the elevations of the two shorelines, which is older?

____

____

____

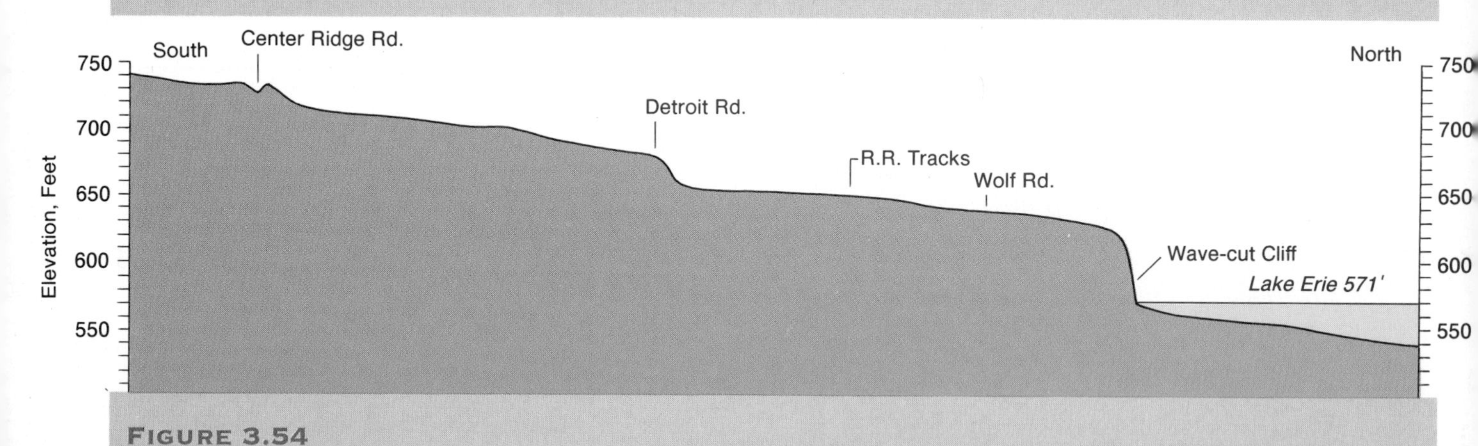

FIGURE 3.54

A generalized north-south topographic profile based on the North Olmstead map (fig. 3.55) showing some ancient shorelines related to the ancestral stages of Lake Erie, Ohio. (See Exercise 18C for exact location of the profile.)

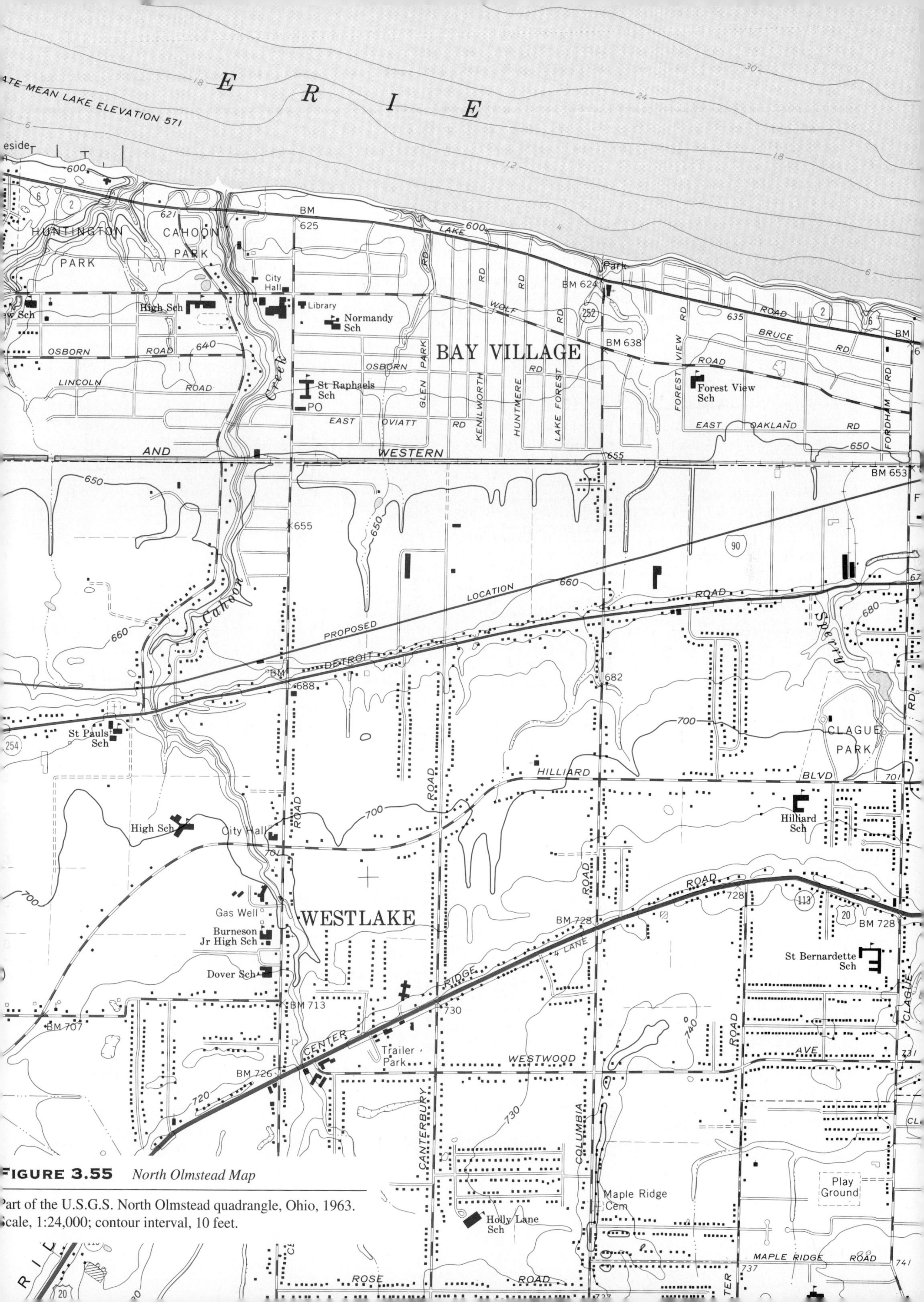

FIGURE 3.55 *North Olmstead Map*

Part of the U.S.G.S. North Olmstead quadrangle, Ohio, 1963. Scale, 1:24,000; contour interval, 10 feet.

Name Section Date

EXERCISE 18D

SHORE EROSION AND LEVELS OF LAKE MICHIGAN

1. The record of levels for Lake Michigan (fig. 3.56) shows highs and lows over a period of 39 years. Does this record suggest that the fluctuations of the lake levels follow a regular periodicity that would permit the forecasting of future lake levels? Explain your answer.

____

____

____

2. What physical characteristics of the sediment exposed in the wave-cut cliff of figure 3.57 made this cliff particularly susceptible to the wave erosion during the high lake levels of 1986?

____

____

____

3. The chart of figure 3.56 shows a pronounced drop in the level of Lake Michigan between 1986 and 1989. Describe the impact of this drop on the sand bluff in figure 3.57.

____

____

____

4. The face of the bluff in figure 3.57 is littered with debris from the abandoned house perched on top. What additional signs are visible on the face of the bluff to indicate that it was undergoing severe erosion at the time the photograph was taken?

____

____

____

____

(Continued)

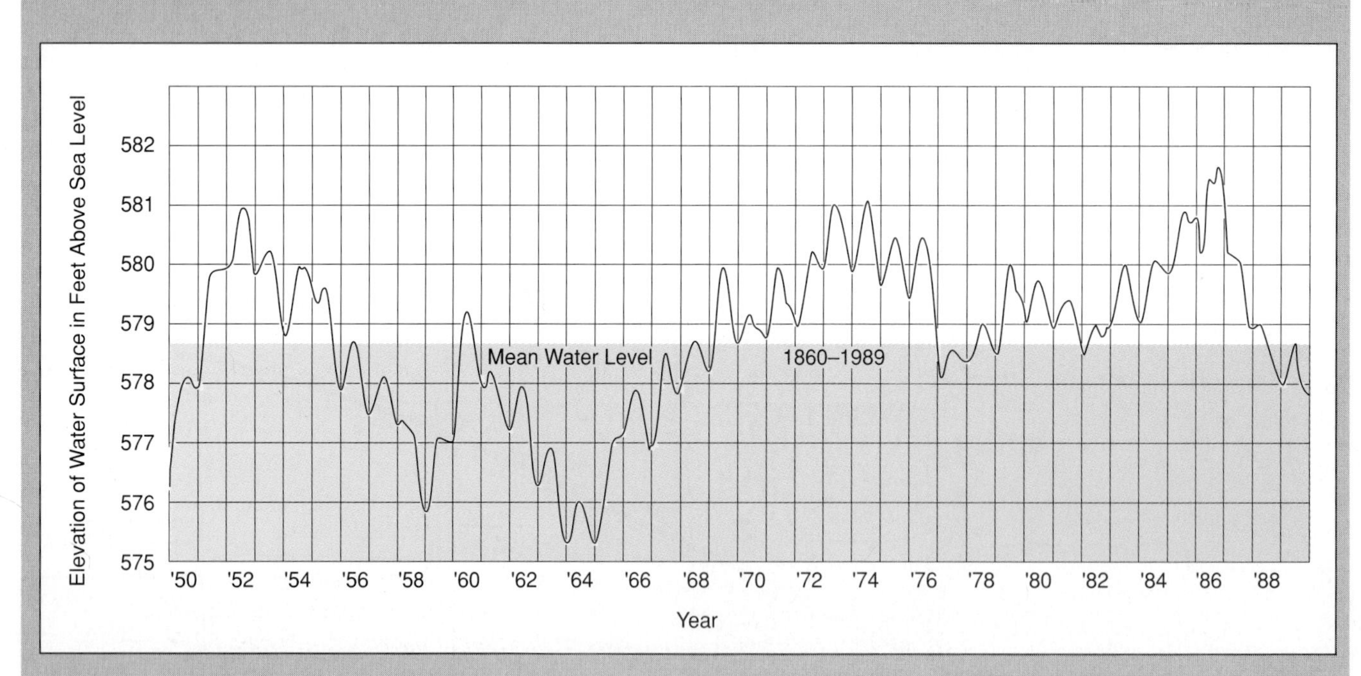

FIGURE 3.56

Water levels of Lake Michigan from 1950 to 1989.

(Based on data published by the U.S. Army Corps of Engineers, Detroit, Michigan 48231.)

FIGURE 3.57

An ancient sand dune exposed to wave erosion on the shore of Lake Michigan near Muskegon, Michigan, November 1986. *(Photo by Marge Beaver, Muskegon, Michigan.)*

Web Connections

http://www.iet.msu.edu/Greatlakes/glhist.htm

Source of interesting information about the Great Lakes including history, lake dynamics, etc.

http://www.lre.usace.army.mil/hmpghh.html

Home page for the U.S. Army Corps of Engineers Great Lakes Hydraulics and Hydrology Branch. A fantastic source of data and images re all of the Great Lakes.

http://www.epa.gov/glnpo

Environmental Protection Agency home page for the Great Lakes National Program Office. Another source of information and data about the Great Lakes.

Name Section Date

EXERCISE 18D CONTINUED

5. What evidence is there in figure 3.58 that an attempt was made to impede if not stop the erosive action of storm waves?

____

____

____

____

6. The cliff in figure 3.58 has receded tens of feet between the time the house was built in the 1930s and the date of the photograph. What happened to the material eroded from the cliff during this period?

____

____

____

____

7. What must happen to the level of Lake Michigan before the stability of cliffs and bluffs along its shore can be restored?

____

____

____

____

FIGURE 3.58

Photograph of a part of the Lake Michigan shoreline near Benton Harbor, Michigan, taken April 16, 1973.
(Hann Photo Service, Hartford, Michigan.)

Landforms Produced by Volcanic Activity

Background

Over 500 active volcanoes occur on earth today, and thousands of extinct volcanoes are spread over the land surface and beneath the oceans. An *active volcano* is one that is in an eruptive phase, has erupted in the recent past, or is likely to erupt in the future. *Dormant volcano* is a term applied to an active volcano during periods of quiescence. An *extinct volcano* is one in which all volcanic activity has ceased permanently.

In this section we will deal with two of the major types of volcanoes, the shield volcano as exemplified by those on the island of Hawaii and the composite or stratovolcano as represented by Mount St. Helens in the state of Washington. A *shield volcano* is a gently sloping dome built of thousands of highly fluid (low viscosity) lava flows of basaltic composition. A *composite volcano* or *stratovolcano* is a conical mountain with steep sides composed of interbedded layers of viscous lava and pyroclastic material. *Pyroclastic* refers to all kinds of clastic particles ejected from a volcano, the most ubiquitous of which is commonly called *volcanic ash.* The lavas in a composite volcano are rhyolite, dacite, or andesite. (Dacite is an aphanitic igneous rock intermediate in composition between rhyolite and andesite.)

Distribution of Volcanoes

Active volcanoes are widespread around the world, but their specific locations are controlled by conditions existing in the earth's crust. Composite volcanoes are concentrated along the margins of tectonic plates that abut each other, and shield volcanoes generally occur over *hot spots,* deep-seated zones of intense heat whose geographic coordinates remain fixed for several millions of years. We will become more familiar with hot spots and tectonic plates in Part 5 of this manual. For purposes of the discussion here, it is sufficient to know that the zone formed where two plate margins converge is characterized by intense volcanic and earthquake activity. The most active convergent plate margin is the "Pacific Rim of Fire," part of which is shown in figure 3.59. Mount St. Helens is one of the many composite volcanoes lying on the North American segment of the Pacific Rim. The hot spot over which the island of Hawaii lies is not associated with a plate margin but occurs beneath the middle of the Pacific plate (fig. 3.59).

The shape and size of a shield volcano and a composite volcano are shown for comparison in the topographic profile of figure 3.60.

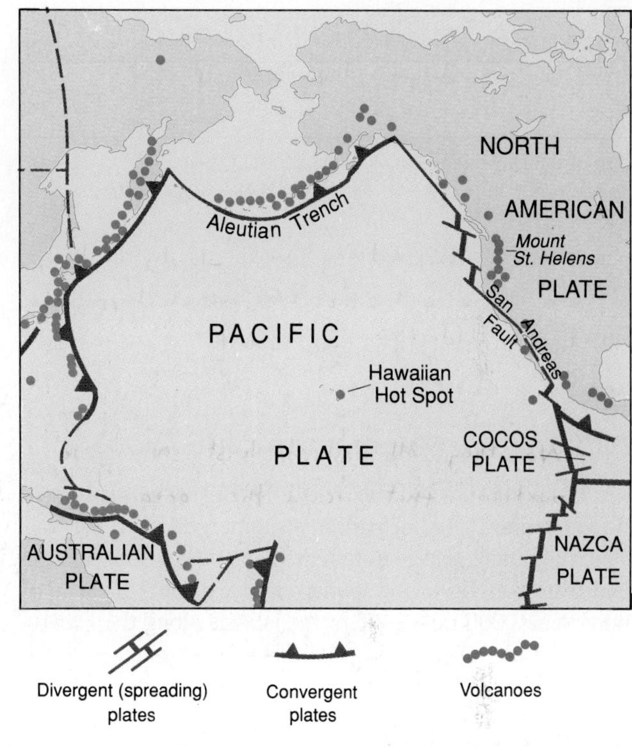

FIGURE 3.59

Map showing locations of the Hawaiian hot spot and Mount St. Helens with respect to part of the Pacific Plate and its margins.
(Source: From R. I. Tilling, et al., 1987. Eruptions of Hawaiian Volcanoes: Past, Present, and Future. U.S. Geological Survey.)

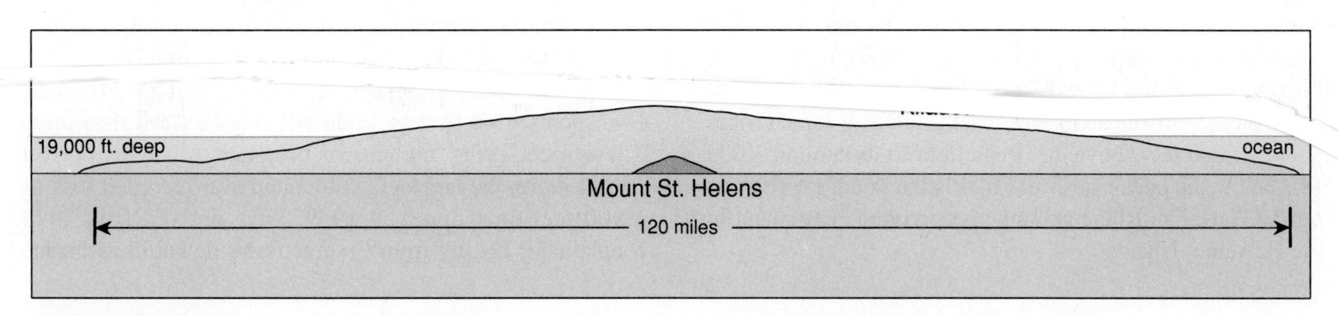

FIGURE 3.60

Topographic profile showing the comparison in size and shape of Mauna Loa and Mount St. Helens.
(Source: From R. I. Tilling, et al., 1987. Eruptions of Hawaiian Volcanoes: Past, Present, and Future. U.S. Geological Survey.)

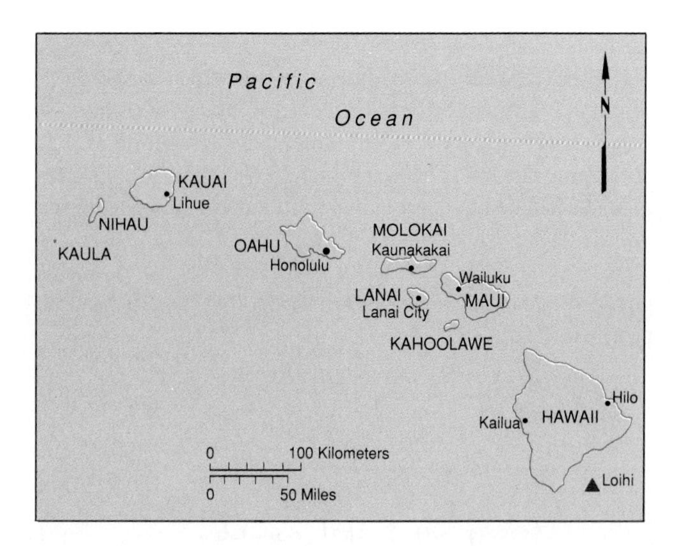

FIGURE 3.61

Map of the Hawaiian Islands in the Pacific Ocean.

(Source: From R. W. Decker, et al., eds., 1987. Volcanism in Hawaii. U.S.G.S. Professional Paper 1350.)

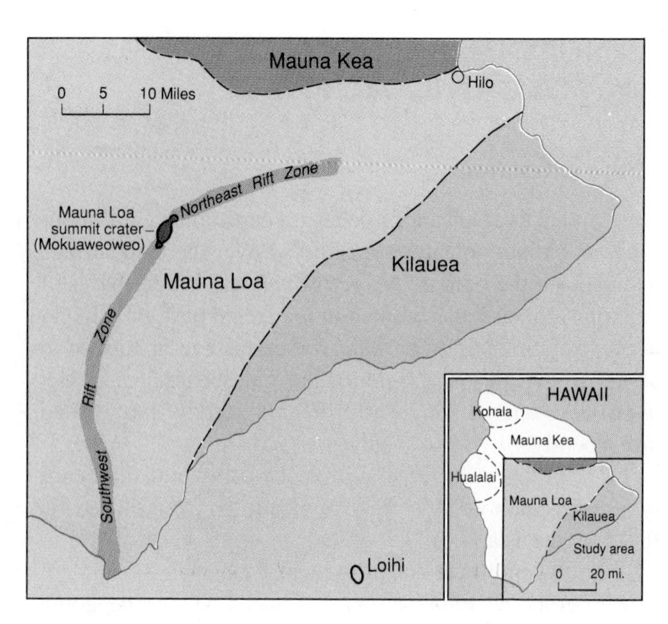

FIGURE 3.62

Map showing the location of the rift zones and the summit crater of Mauna Loa on the Island of Hawaii. Inset shows the names and boundaries of the five shield volcanoes that form the architecture of the Big Island.

(Source: Modified from R. I. Tilling, et al., 1987. Eruptions of Hawaiian Volcanoes: Past, Present, and Future. U.S. Geological Survey.)

The Shield Volcanoes of Hawaii

The Island of Hawaii, also called the Big Island, is part of the Hawaiian Ridge, a partially submerged chain of volcanic mountains that rise from the deep ocean floor and extend from the Big Island to the northwest for a distance of more than 2,000 miles. All of the islands along the Hawaiian Ridge are shield volcanoes that were formed over the Hawaiian hot spot as the ocean floor moved progressively to the northwest over the last 40 million years.

The Big Island of Hawaii is the largest of the Hawaiian Islands (fig. 3.61) and consists of five shield volcanoes, of which only two, Mauna Loa and Kilauea, are active (fig. 3.62). Together, these five volcanoes rise 30,000 feet above the floor of the Pacific Ocean to an elevation more than 13,000 feet above sea level. The oldest of the five is Kohala, which became inactive about 60,000 years ago; Mauna Kea, the next oldest, ceased its eruptive activity about 3,000 years ago. Hualalai has erupted only once in historic times, during 1800–1801. Loihi, an active submarine volcano off the southeast coast of the Big Island, rises about 10,000 feet above the ocean floor to its summit 3,000 feet below the ocean surface. Loihi is the youngest volcano on the Hawaiian Ridge and may become the next island in the Hawaiian Islands.

Hawaiian-Type Eruptions

Almost all Hawaiian-type volcanic eruptions on the Big Island during historic times (roughly the period from 1843 to the present) have occurred on Mauna Loa and Kilauea. These two volcanoes have been intensely studied by volcanologists in recent years, and it is from their observations that we are able to know a great deal about Hawaiian-type eruptive activity.

Hawaiian-type eruptions are weakly explosive or non-explosive. They extrude highly fluid basaltic lava that flows easily down the gentle slopes of the volcano's flanks. Indeed, it is the low viscosity of the lavas that accounts for the gentle slopes to begin with. Lava is erupted not only from the crest of the volcano but also from rift zones along its flanks. The rift zones on Mauna Loa are called the Southwest Rift Zone and the Northeast Rift Zone (fig. 3.62). The lava extruded from these rifts comes from a magma reservoir a few miles beneath the summit.

In the early stages of a Hawaiian-type eruption, lava spouts from fissures in the rift zones as *lava fountains.* These spectacular "curtains of fire" rise to heights of 100 feet or more; the largest lava fountain ever recorded rose to 1,900 feet on Kilauea in 1959. Lava derived from lava fountains or oozing from fissures flows downhill as incan-

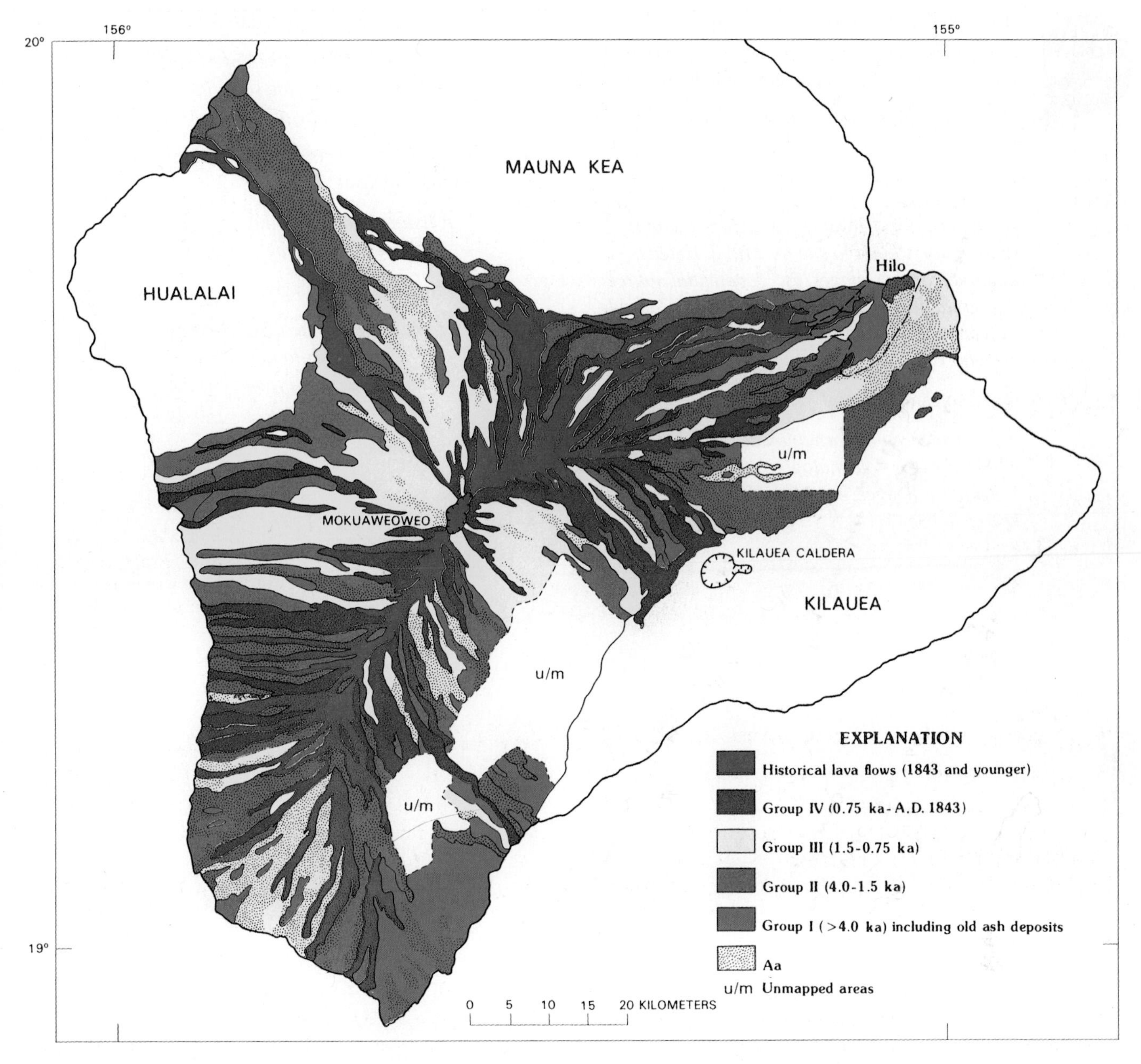

FIGURE 3.64

Map of Mauna Loa showing the surface distribution of lava flows in five different age categories. The notation "ka" stands for thousands of years before the year 1950. Thus, 0.75 ka = 750 years; 1.5 ka = 1,500 years; and 4.0 ka = 4,000 years.

(From J. P. Lockwood and P. W. Lipman, 1987, Holocene Eruption History of Mauna Loa Volcano, Chapter 18 in R. W. Decker et al., editors, Volcanism in Hawaii, U.S. Geological Survey Professional Paper 1350.)

WEB CONNECTIONS

http://www.soest.hawaii.edu/mauna_loa/

Home page reporting research on the world's largest volcano.

http://vulcan.wr.usgs.gov/home.html

Home page for the USGS Cascades Volcano Observatory. Links to site for Mt. St. Helens, other volcanoes, images, etc. A potential source of a great deal of information about the Cascade Volcanoes of British Columbia, Washington, Oregon, and California.

http://volcano.und.edu/

A great source of information about volcanoes. Click on Ask a Volcanologist, currently Erupting Volcanoes, Images, etc. More than enough to keep you busy here.

http://www.geo.mtu.edu/volcanoes/

A web page maintained by Michigan Tech University to provide scientific and educational information to the public about volcanoes.

http://www.usgs.gov/network/science/earth/volcano.html

A listing of sites dealing specifically with volcanoes around the world.

http://hvo.wr.usgs.gov/

Home page for the USGS Hawaiian Volcano Observatory. Up-to-date information on volcanic activity, past history, and hazards.

The Eruption of Mount St. Helens

On May 18, 1980, Mount St. Helens, one of the several spectacular volcanic peaks in the Cascade Range of the Pacific Northwest, underwent a violent volcanic eruption after two months of low-level activity characterized by earthquakes, steam venting, and small ash eruptions. This eruption of Mount St. Helens, ending a 123-year dormant period, began with seismic activity on March 20, 1980. Steam vents opened up and ash began to be erupted by March 27. Small craters close to the summit developed, ash plumes were erupted, and ash avalanches took place in early April. Seismic activity continued, and by mid-April a significant crater had formed. The summit area of Mount St. Helens began to swell, enough so that Goat Rocks on the northern flank had a measured movement of 20 feet vertically and 9 feet horizontally to the northwest.

As this activity continued, ash and steam continued to be produced, but in late April seismic activity was greatly reduced. In early May as swelling of the peak continued, the U.S.G.S. reported that the northern rim of the crater was rising at a rate of 2 to 4 feet per day.

Through the use of remote sensing techniques utilizing infrared film, hot spots were recorded in early May. Swelling continued, seismic activity increased, and at 8:32 A.M. on May 18, 1980, the violent eruption of Mount St. Helens occurred, an event that claimed over 60 lives, devastated over 200 square miles of timberland and recreational areas, and spread measurable thickness of ash over several states and, eventually, around the world.

The eruption itself was marked initially by an earthquake that triggered a major landslide down the north side of the mountain, followed quickly by a violent explosion. The two events combined to destroy the north rim of the summit crater (location 7, fig. 3.66). The rock material moved down the north side of the mountain as a mixture of rock, ash, steam, and glacial ice (location 6, fig. 3.66). Additional fluidization occurred when this avalanche mass hit the water of Spirit Lake and Toutle River. The velocity of the avalanche has been estimated at over 150 mph.

A portion of this enormous avalanche flowed down the valley of the Toutle River for 13 miles, depositing materials in a swath up to 1.2 miles wide and with a thickness up to 450 feet (fig. 3.66). Another part of the avalanche continued to the north, rose over a ridge that was 1,000 feet high, depositing over 100 feet of debris on top of the ridge before pouring over into the valley of South Coldwater Creek on the north side of the ridge (location 4, fig. 3.66). New lakes were formed as stream valleys were dammed by debris (location 3, fig. 3.66) and new islands were formed in Spirit Lake (location 5, fig. 3.66). The heavy black line on figure 3.66 marks the southern edge of the "Eruption Impact Area" as defined by the U.S. Geological Survey.

Large areas were covered by mudflows (unstippled gray areas, fig. 3.66) that resulted from the mixture of water from melting glaciers and large quantities of ash that poured out during the eruption. The major river draining the area to the north of Mount St. Helens, the Toutle River, carried large quantities of sediment almost as mudflow to the west into the Cowlitz River and eventually into the Columbia River. Previously recorded flood stages on the Toutle River were exceeded by almost 30 feet. Silting occurred in the Columbia River at the mouth of the Cowlitz River, trapping ships upstream. Dredging of a channel was necessary before these ships could move downriver to the ocean.

Ash falls occurred to the east of Mount St. Helens affecting cities such as Yakima, in the heart of the apple growing district of Washington, and the major wheat growing areas farther east. At Ritzville, 205 miles east of Mount St. Helens, a fine ash deposit of 70 mm was recorded, and by May 21 ash had spread across the continent to the east coast. A later ash eruption on May 25 spread ash to the northwest, mantling an area lying roughly between the Columbia River and Olympia, Washington. Small ash eruptions have occurred since, but none as large as the May 25 activity.

The eruption of Mount St. Helens was the first such volcanic event in the contiguous 48 states since the eruption of Mt. Lassen in northern California that began in 1914 and continued until 1921. Volcanic activity in the Cascade Range was recorded during the 1800s, and several peaks such as Mt. Rainier, Mt. Baker, and Mt. Hood still have active fumaroles.

Volcanic activity has continued on Mount St. Helens at a much reduced rate in the years following the 1980 eruption. This activity has included gas and ash emissions, earthquakes, rockfalls, and extrusion of a lava spine in the crater, and there is now forming a "composite dome" in the center of the crater. This activity is continually monitored by the U.S.G.S. and is providing considerable information that may help in the development of earthquake and volcanic eruption prediction models.

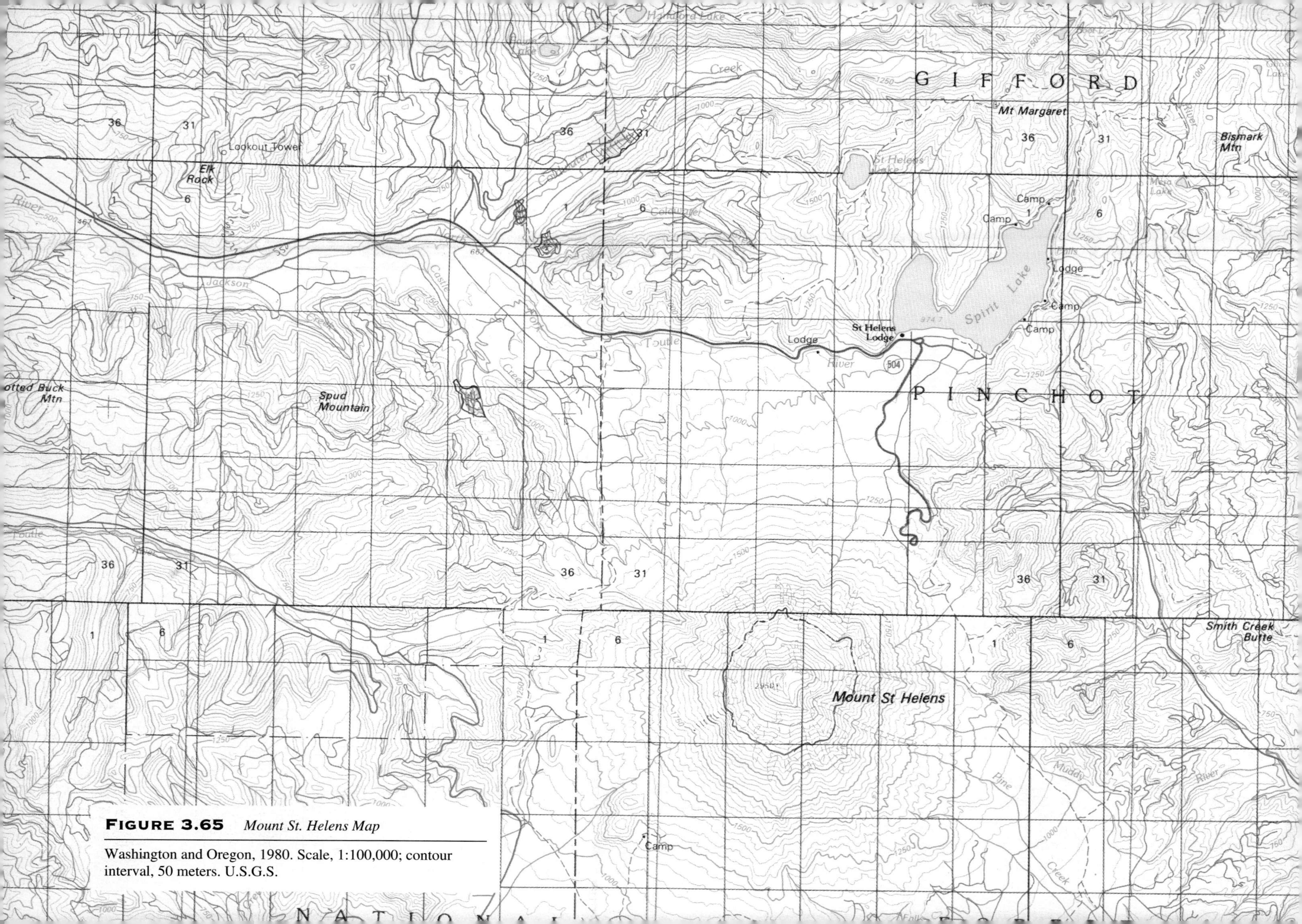

FIGURE 3.65 *Mount St. Helens Map*

Washington and Oregon, 1980. Scale, 1:100,000; contour interval, 50 meters. U.S.G.S.

PART FOUR
STRUCTURAL GEOLOGY

BACKGROUND

Structural geology deals with the architectural patterns of rock masses found in nature. In Part One you were introduced to some basic concepts and principles about the occurrence of rocks. In Part Four, we will build on the concepts and related terminology introduced in Part One. It may be useful for you to review the text and diagrams on pages 55–62.

Structural geology involves all three rock types—igneous, sedimentary, and metamorphic—but in this part of the manual we concentrate on structures in which sedimentary rock layers are dominant. A sedimentary rock unit that is characterized by a distinct lithologic composition is called a *formation.* A formation is the basic stratigraphic unit depicted on a geologic map. The boundary between two contiguous formations is called a *contact.*

For the sake of simplicity, the formations illustrated herein will be homogeneous in their lithology and will be referred to by some name such as "limestone formation," "sandstone formation," or some other name. In reality, however, a formation may consist of several thinner layers or beds. The contacts between these beds are called *bedding planes* and are more or less parallel to the contacts of the formation itself.

It is common geologic practice to assign a name to a formation. In some cases, the name of the formation will include a litholog descriptor such as the Madison Limestone or the Pierre Shale. In other cases, the name will consist only of a proper name such as the Nelson Formation or the Sunflower Formation. Formational names such as these will be found only on published geologic maps. On simple diagrammatic maps that are used herein, it will suffice to use only a lithologic definition such as "shale formation" or "sandstone formation."

Sedimentary strata occur in a variety of three-dimensional geometric forms. The pattern of contacts between sedimentary formations portrayed on a map shows the distribution of the formations in only two dimensions. The visualization of a three-dimensional geometric form from a two-dimensional geologic map is one of the main objectives of this part of the manual. By learning certain basic principles and following standard procedures described in the pages that follow, the ability to formulate a mental three-dimensional picture of the geometric configuration of strata beneath the earth's surface can be mastered.

Geologic structures are produced when strata are deformed by forces that contort the strata from their original position of horizontality into geometric forms called *folds.* Folded rock layers retain their continuity as layers, that is, the contact between two formations on a geologic map can be followed as an unbroken line. When exposed by erosion, these folds are revealed in a two-dimensional *outcrop pattern* on a geologic map that is diagnostic of the three-dimensional forms of the structures.

In many cases, deformation causes the strata to break or *fault* along *fault planes* so that the outcrop pattern shows a discontinuity of the formations on either side of the fault. Movement along a fault plane produces an *earthquake.*

Folds, faults, and earthquakes therefore are the subject of Part Four and will be treated in that order in the pages that follow.

Structural Features of Sedimentary Rocks

Deformation of Sedimentary Strata

During the course of geologic history, sedimentary strata have been subjected to vertical and horizontal forces that may alter the original horizontal position of the rock layers. Some strata may be uplifted in a vertical direction only, so that their original horizontality remains more or less intact. In other cases, the forces of deformation produce architectural patterns ranging from simple to extremely complex structures.

In order to decipher these structures, geologists measure certain features of a given formation where it crops out at the surface of the earth. These measurements define the position of the formation with respect to a horizontal plane of reference. The precise orientation of a contact, bedding plane, or any planar feature associated with a rock mass is called the *attitude.* When attitudes from many outcrops are plotted on a base map, such as a topographic map or aerial photograph, and combined with the contacts between formations, the overall geometric pattern or structural configuration of the strata can be determined.

Components of Attitude

The attitude of a formation consists of three parts that collectively define its position at a given location with respect to a horizontal plane and a compass direction. The three components of attitude are shown in figure 4.1.

1. *Strike:* A horizontal line in the plane of the bedding expressed as a compass direction.
2. *Direction of dip:* The compass direction in which the layer is inclined downward from the horizontal. The direction of dip is always at right angles to the direction of the strike.
3. *Dip angle:* The angle between a horizontal plane and a bedding plane. The dip angle is measured in degrees.

As an example of a verbal description of the attitude of a formation at a particular site, the following notation would be used: At the north end of the bridge across the Snake River, the Sundance Formation strikes north 45 degrees east and dips to the southeast at an angle of 30 degrees. On a geologic map, however, the attitude of a formation would be shown by a *strike and dip symbol.* Various forms of this symbol are given in figure 4.2. In illustrations used in parts of this manual, the strike and dip symbols may appear without the notation of the angle of dip.

Methods of Geologic Illustration

Geologic information, gathered by the study of outcrops at the surface and through the use of subsurface information obtained from wells, is displayed in a number of ways in order to depict the overall structural features and relative age relations of the strata involved. The three main types of geologic illustrations or diagrams and their relationships are shown in figure 4.3. Note that there is no surface expression of the limestone unit. This subsurface information came from local wells.

1. *Geologic map:* This is a map that shows the distribution of geologic formations. Contacts between formations appear as lines, and the formations themselves are differentiated by various colors. The map may also show topography by standard contour lines.
2. *Geologic cross section:* A diagram in which the geologic formations and other pertinent geologic information are shown in a vertical section. It may also show a topographic profile, or it may be schematic and show a flat ground surface.

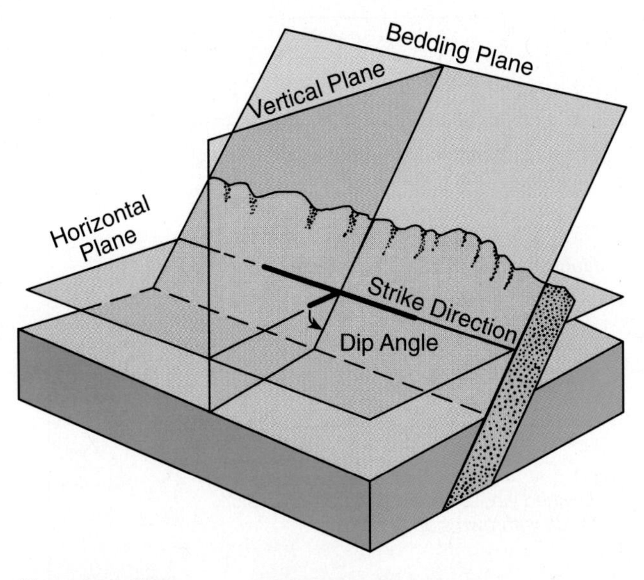

FIGURE 4.1

Three-dimensional view of an outcropping of sandstone in which the *attitude* of a bedding plane is measured with respect to horizontal and vertical planes. The shaded slanting plane represents the bedding plane of the layered sandstone. The intersection of the bedding plane and a horizontal plane results in a line called the *strike* of the formation. On a map this line is expressed as a compass direction. The angle formed by the horizontal plane and the bedding plane is the *dip* of the formation. The *dip angle* is always measured in a vertical plane that is perpendicular to the direction of strike.

(After Elements of Physical Geology, J. H. Zumberge and C. A. Nelson. Copyright © 1976 John Wiley & Sons, Inc.)

3. *Block diagram:* A perspective drawing in which the information on a geologic map and geologic cross section are combined. This mode of geologic illustration is used to show the three-dimensional aspects of a geologic structure.

SEDIMENTARY ROCK STRUCTURES

Sedimentary strata that have been subjected to forces of deformation result in three basic geologic structures as shown in the block diagrams of figure 4.4.

A. *Monocline:* A one-limb flexure (fold) in which the strata have a uniform direction of strike but a variable angle of dip.
B. *Anticline:* A fold, generally convex upward, whose core contains the stratigraphically older rocks.
C. *Syncline:* A fold, generally concave upward, whose core contains the stratigraphically younger rocks.

Notice that in figure 4.4 the arch of the anticline is not reflected in a corresponding topographic arch and that the synclinal trough is a geologic trough, not a topographic one. The surface topography of the parallel ridges in figures 4.4B and 4.4C is controlled by a formation that is more resistant to erosion than the other formations in the structure.

GEOMETRY OF FOLDS

The geometry of a fold is more precisely defined by the attitude of the *axial plane* of the fold, an imaginary plane that separates the *limbs* of the folds into two parts (as symmetrically as possible) as shown in figure 4.5A. The *axial trace* of the fold appears as a line on a geologic map.

If the axial plane is essentially vertical, the fold is said to be *symmetric* (fig. 4.6A); if the axial plane is inclined so that the limbs dip in opposite directions but one limb is steeper than the other, the fold is *asymmetric* (fig. 4.6B); and if the axial plane is inclined to the extent that the opposite limbs dip in the same direction, the fold is *overturned* (fig. 4.6C). A *recumbent fold* is an overturned fold in which the axial plane is nearly horizontal. The symbols used on geologic maps to show the traces of axial planes are shown in figure 4.2.

The folds shown in figures 4.4, 4.5, and 4.6 are *nonplunging folds* because the strikes of the limbs are parallel. Another way of describing a nonplunging fold is to say that the strikes of the folded formations are all parallel, as shown in figure 4.3.

If, however, the strikes of the formations on either side of an axial plane converge, as in figure 4.7, the fold is said to be a *plunging fold.* A geologic map on which a series of plunging folds is displayed shows a *zig-zag* outcrop pattern. The *direction of plunge* is shown by an arrow placed on the trace of the axial plane, as in figure 4.7. *The direction of plunge of a plunging anticline is toward the apex of the converging formations* as seen on a geologic map, and *the direction of plunge of a plunging syncline is toward the open end of the V-shaped pattern of diverging formations.* Figure 4.7 shows both cases.

An anticline that plunges in opposite directions is a *doubly plunging anticline,* and a syncline that plunges in opposite directions is a *doubly plunging syncline.* Variations of doubly plunging folds are the *structural dome* and *structural basin,* as shown in figure 4.8. The outcrop patterns of these two structures are more or less concentric circles.

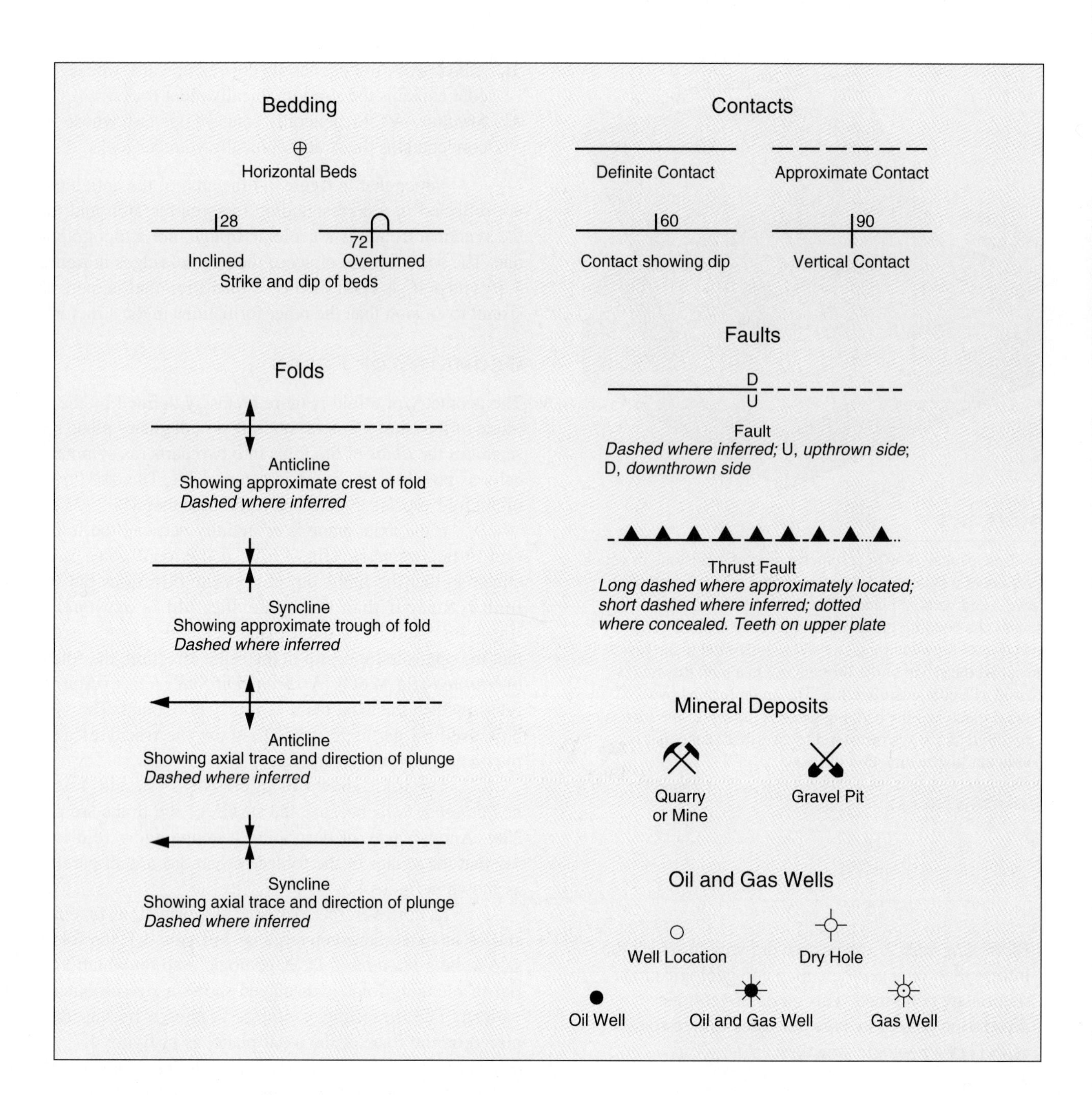

FIGURE 4.2

Standard symbols used on geologic maps.

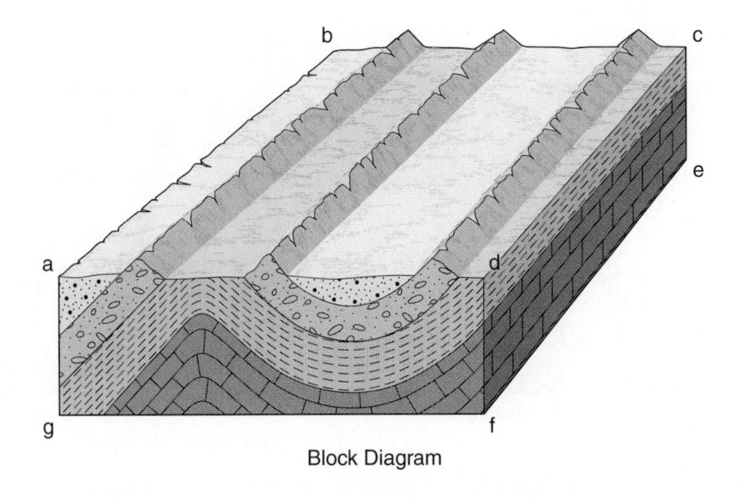

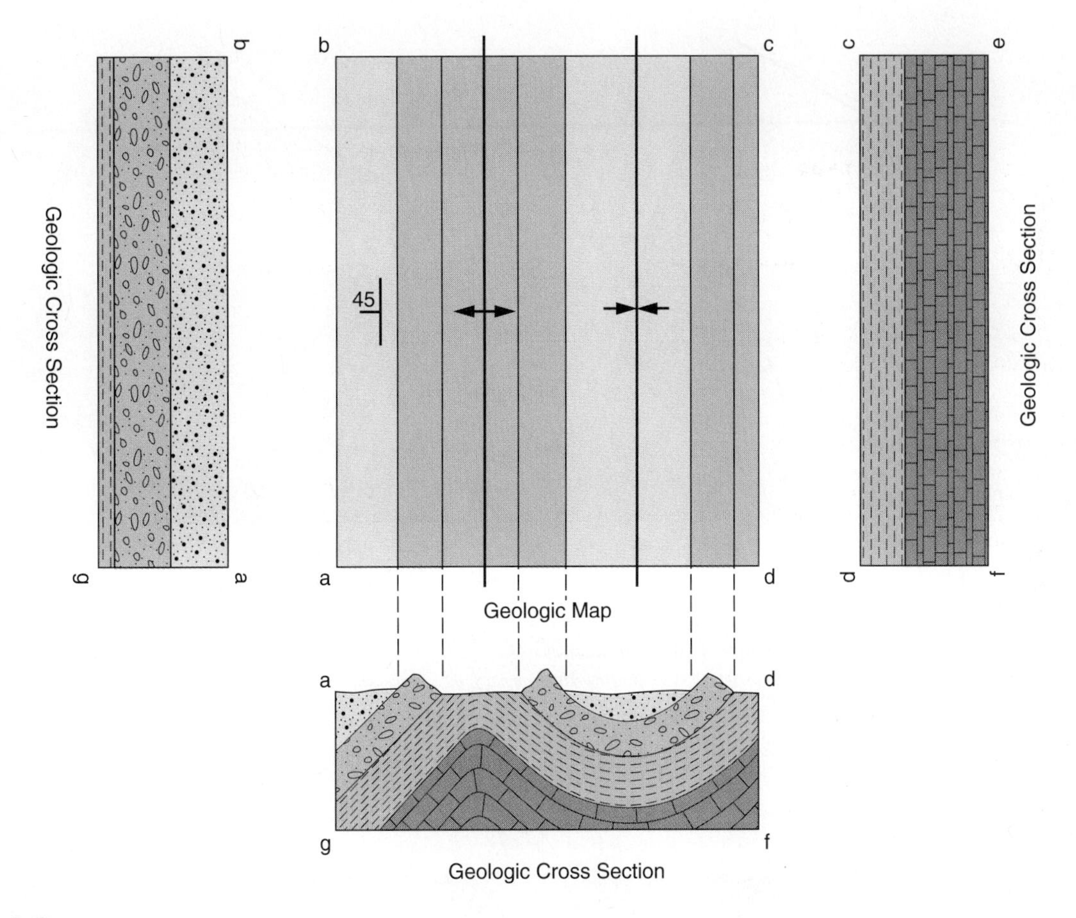

FIGURE 4.3

The *block diagram* is a three-dimensional drawing in which the geometric configuration of the geologic structure is depicted. A *geologic map* shows the areal extent of formations at the earth's surface and contains certain symbols that further define the geometry of the rock masses as they extend beneath the surface. A *geologic cross* section is a view of the geologic formations in a vertical plane.

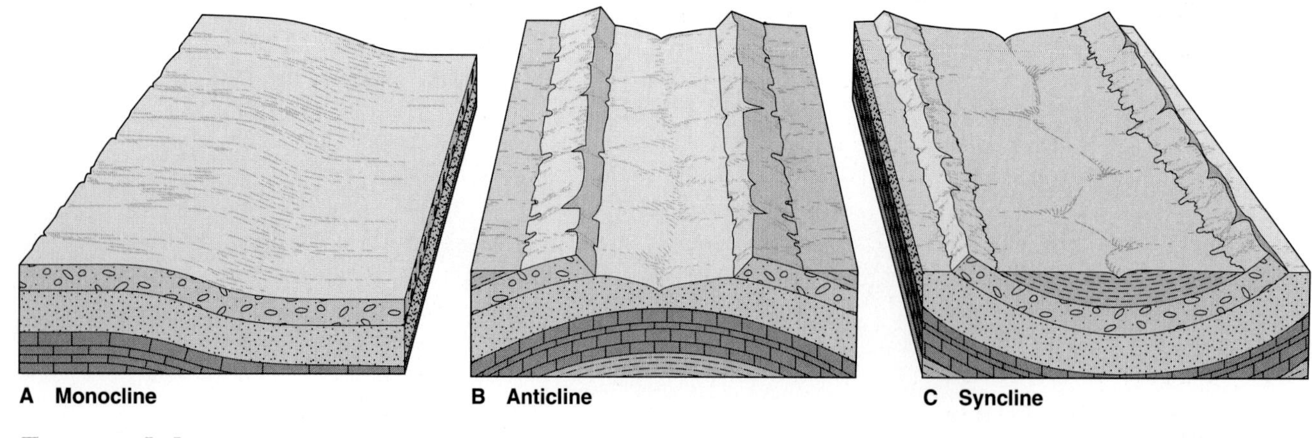

FIGURE 4.4

Block diagrams of three common folds. The dissected ridges formed by resistant layers in diagrams *B* and *C* are called hogback ridges.

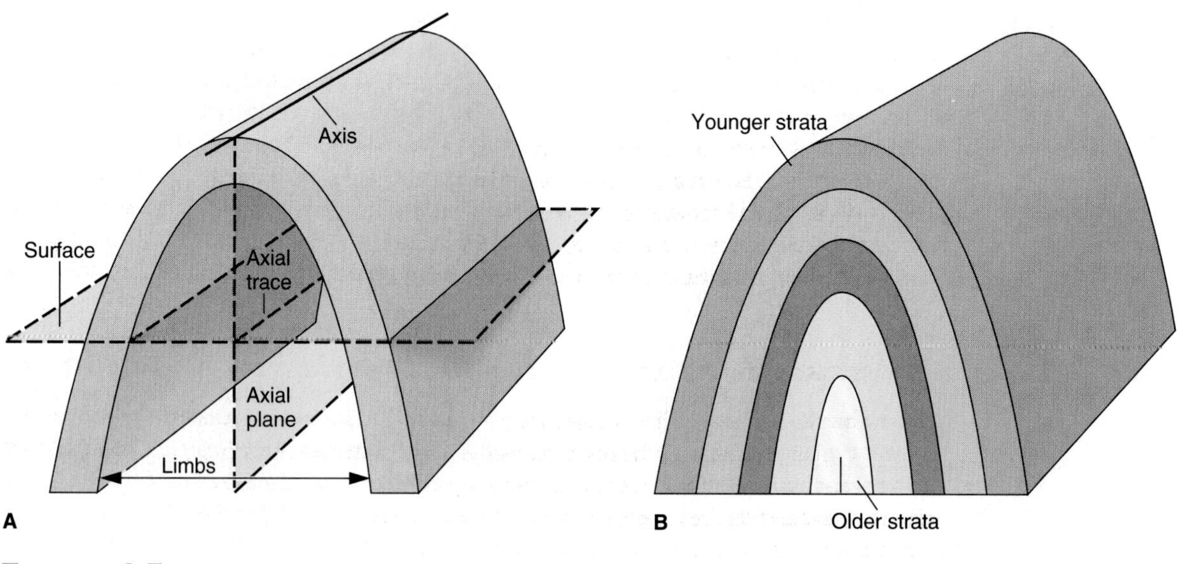

FIGURE 4.5

(*A*) Nomenclature of a fold. (*B*) Age relationships of strata in an anticline.

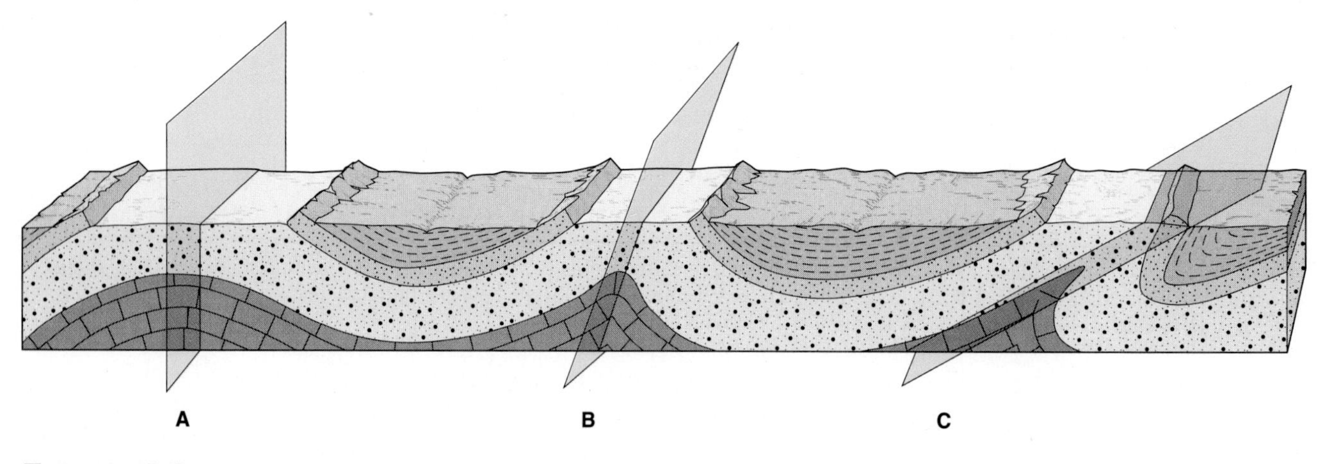

FIGURE 4.6

Block diagram in which three variations of a fold are shown: (*A*) symmetric anticline; (*B*) asymmetric anticline; (*C*) overturned anticline. Note the different attitudes of the three axial planes.

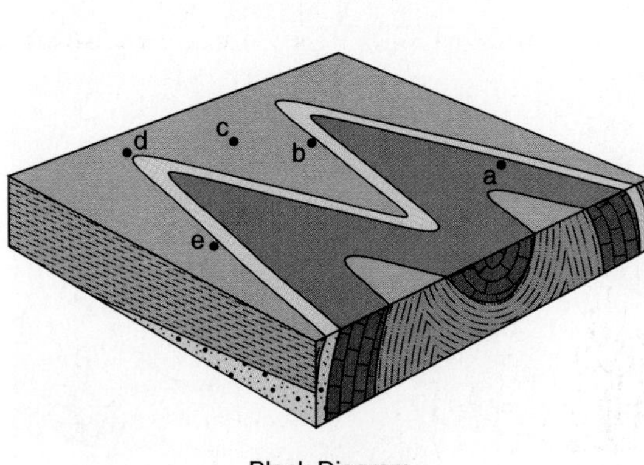

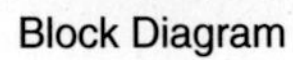
Block Diagram

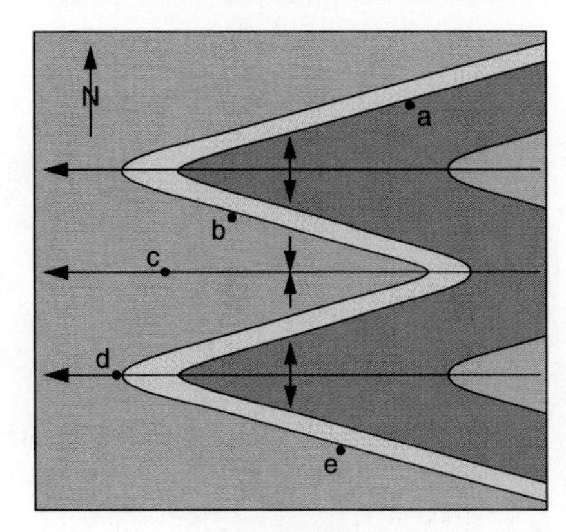

Geologic Map

FIGURE 4.7

Block diagram and geologic map of plunging folds. The map shows the characteristic outcrop pattern of plunging folds. Here, two anticlines and one syncline plunge to the west. If this area were in a humid region, the arkose formation would be more resistant to erosion than the shale, limestone, or siltstone formations and would therefore form a hogback ridge.

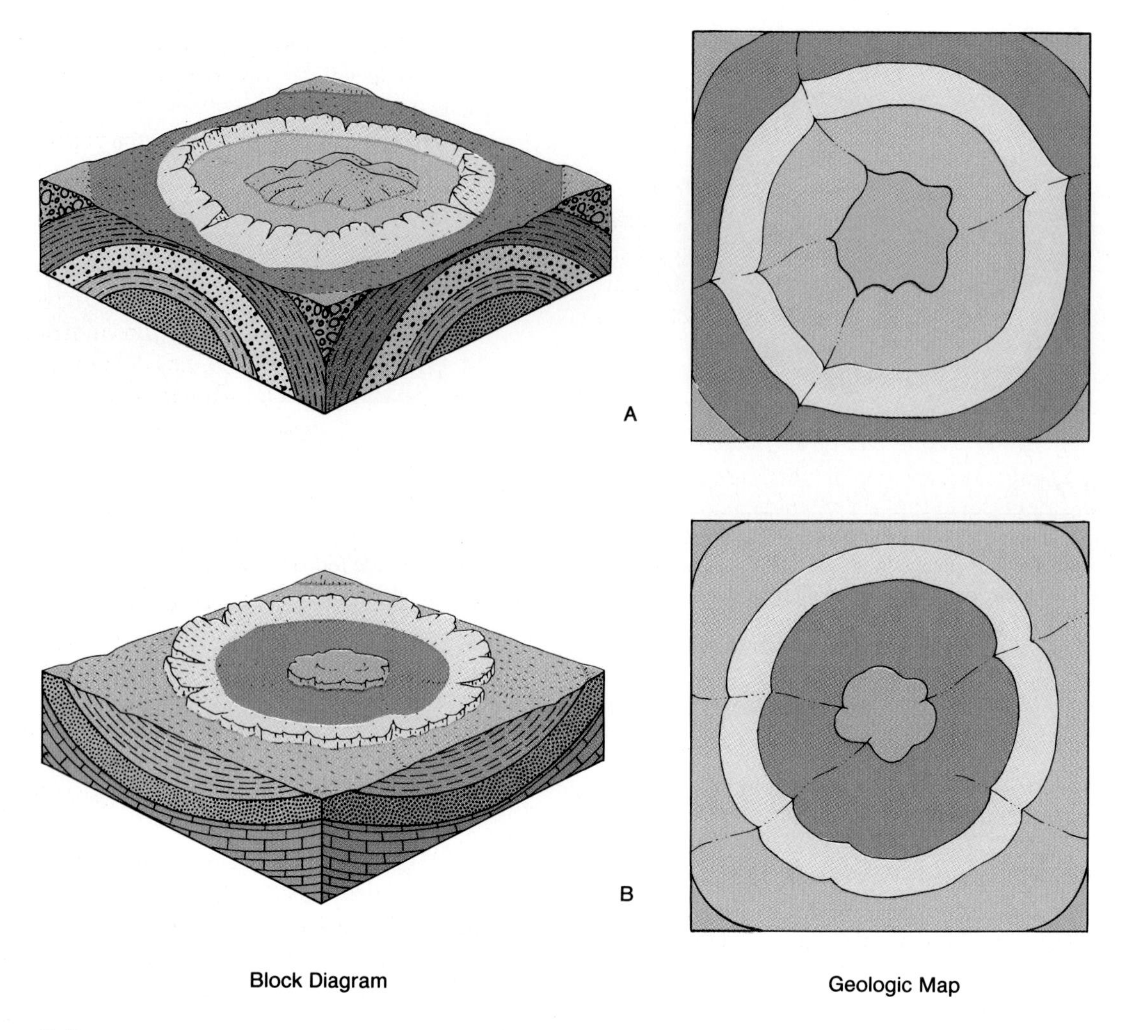

FIGURE 4.8

Block diagrams and geologic maps of two structures that produce circular outcrop patterns: (*A*) structural dome; (*B*) structural basin.

Geologic Maps and Cross Sections

Geologic Maps

A geologic map shows the distribution of rock types in an area. The map is constructed by plotting strikes and dips of formations and the contacts between formations on a base map or aerial photograph. This information is based on field observations on outcrops in the map area. Because a single isolated outcrop rarely yields sufficient information from which the overall structural pattern for a given area can be understood, the geologist must visit enough outcrops in the map area to permit the filling of the gaps from one outcrop to another.

We are not concerned here with the making of actual geologic maps but rather with their interpretation and the construction of geologic cross sections from them. The interpretation of a geologic map requires an understanding of the information shown on it and the ability to translate that information onto a geologic cross section. To accomplish this, one must learn to visualize three-dimensional relationships from a two-dimensional pattern of geologic formations as they appear on a geologic map. This is perhaps the most difficult aspect of structural geology for a beginning student to master, but by following step-by-step instructions, these relationships eventually will become clear.

In general, keep in mind that a geologic map shows the distribution of formations as they appear at the surface of the earth. How this surface information can be used to visualize the unseen components of the rocks below the surface constitutes the subject matter to follow.

Having already been exposed to the notations and symbols used on a geologic map, you may find it useful to review the relationship between a geologic map and a geologic cross section as shown in figure 4.3. If you thoroughly understand how the geologic cross section of figure 4.3 relates to the geologic map and how both relate to the block diagram, you will be in a good position to proceed with the next steps in map interpretation.

Geologic Cross Sections

The purpose of a geologic cross section is to display geologic features in a vertical section, perpendicular to the ground surface. A crude but nonetheless accurate analogy is to be found in the ordinary layer cake. When a layer cake is viewed from above, all that can be seen is the frosting; the "structure" of the cake is obscured. However, if the cake is cut vertically and the two halves are separated, the component layers of the cake constitute a cross section of the cake so that its structure will be revealed.

Geologic formations do occur in "layer-cake" structures, but they commonly occur in much more complex structures, and it is through the construction of a geologic cross section that these complexities are unraveled. Following are some general rules and guidelines for use in constructing a geologic cross section from a geologic map.

1. A geologic cross section is constructed on a vertical plane. The cross section is shown on the corresponding geologic map by a line that is equivalent to the line along which the cake was cut in the layer-cake analogy. Information on or near the line of the cross section on the map is transferred to the cross section as the first step in its construction. Such notations as directions and angles of dip, formational contacts, traces of axial planes, and the like provide the basic elements used to make a geologic cross section.
2. Sedimentary formations to be drawn on cross sections in the exercises in this manual are assumed to have a constant thickness. That is to say, they do not thicken or thin with depth or along the strike.
3. Dip angles from strike and dip symbols on the map can be used as a basis for estimating the inclination of strata on a cross section. If dip angles are not shown, keep the dip angles as small as possible but consistent with the thickness of the strata and structural relationships.
4. The relative ages of sedimentary strata in some of the maps and cross sections used herein are designated by arabic numerals. For example, if four formations are shown on a map or block diagram, the oldest formation is assigned the number "1," and the youngest, a number "4" (fig. 4.9A).
5. If you are required to draw a geologic cross section from a geologic map on which no strike and dip symbols are present, the direction of dip can be determined in the following manner.
   (*a*) Where a formation contact crosses a stream on the map, it forms a *V*, the apex of which points in the direction of dip as shown in the geologic map of figures 4.9*B* and 4.9*C*. (This rule is not to be confused with the "law of V's" as applied to contour lines when they cross a stream.)
   (*b*) The shape of a *V* formed by a contact that crosses a stream may be used to estimate the angle of dip of the contact. A broad open *V* indicates a steep dip, and a narrow *V* is indicative of a shallow dip angle. Where no *V* is formed, the formation contact is vertical, as shown in figure 4.9*D*. The foregoing method for determining the direction of dip takes precedence over the method described next.
   (*c*) In a sequence of formations, none of which has been overturned, the oldest beds dip toward the youngest, as shown in the geologic map of figures 4.9*B* and 4.9*C*.

Width of Outcrop

When folded strata are exposed to erosion at the earth's surface, they appear as bands on the geologic map. The width of a single band is called the *width of outcrop* although the full thickness of the formation may not be exposed in a single outcrop. The width of outcrop is controlled by three factors: the thickness of the formation, the angle of dip of the formation, and the slope of the land surface where the outcrop is exposed.

To illustrate these controlling factors in the simplest case, consider the three horizontal formations of equal thickness in figure 4.10. The geologic cross section in figure 4.10*A* shows how the thickness of each formation varies with the slope of the land surface. A gentle slope results in a width of outcrop that is greater than the thickness of the formation, as in the case of the shale formation; and a steeper slope produces a width of outcrop that is less than the thickness of the formation, as in the case of the sandstone and limestone formations.

Two other cases of the relationship of thickness to the width of outcrop are shown in figure 4.10*B* and *C*. In figure 4.10*B*, where the beds are dipping 30 degrees, the thickness of each formation is shown on the cross section and the corresponding width of outcrop is shown on the geologic map. In figure 4.10*C*, the formations are vertical, that is, they dip 90 degrees. In this case, the true thickness of a formation is the same as the width of outcrop. The general rule, however, is that *the width of outcrop on a geologic map is not necessarily the same as the true thickness of the formation as seen in a geologic cross section.* This rule must be kept in mind when drawing cross sections from a geologic map or a block diagram in the exercise that follows.

FIGURE 4.9

Block diagram and maps showing the relationship of topography to outcrop patterns. In all cases the stream flows from north to south.

(A) Horizontal strata dissected by a drainage system. Numbers refer to relative ages of the formations. The formation labeled 1 is the oldest. The apex of the V formed will point upstream and will be parallel to the contours.

(B) Tilted rock strata dipping downstream at an angle steeper than the stream channel. The oldest beds (i.e., 1, 2, and 3) dip toward the youngest beds (5 and 6). The apex of the V formed points downstream and in the direction of dip.

(C) Titled strata dipping upstream. The apex of the V formed points upstream and in the direction of dip but the contact crosses contours.

(D) Vertical sedimentary beds, one of which is more resistant to erosion than the other two. In this case the law of V's cannot be used because no V's are formed. Thus, the age relationship cannot be determined from the information either on the block diagram or on the map.

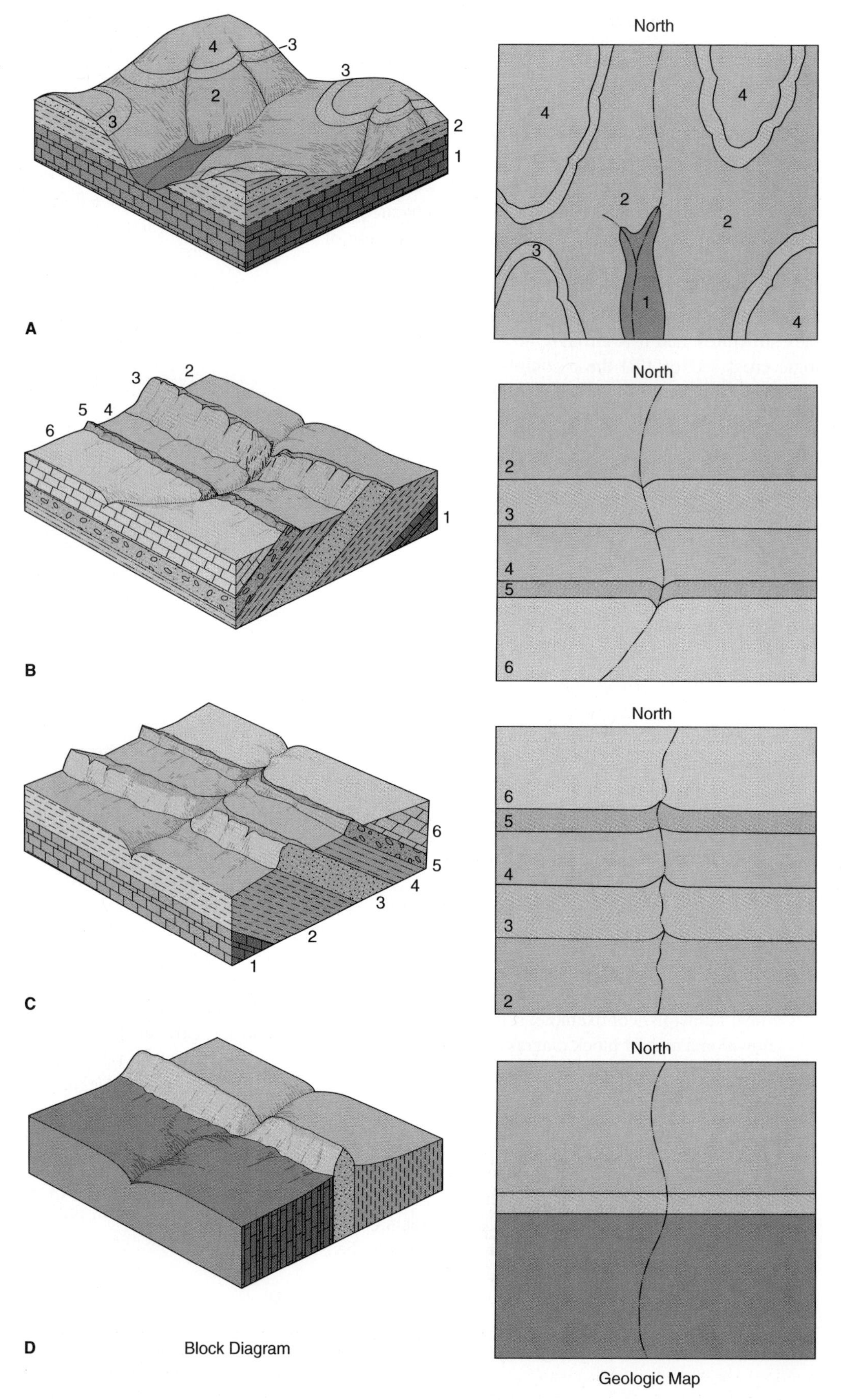

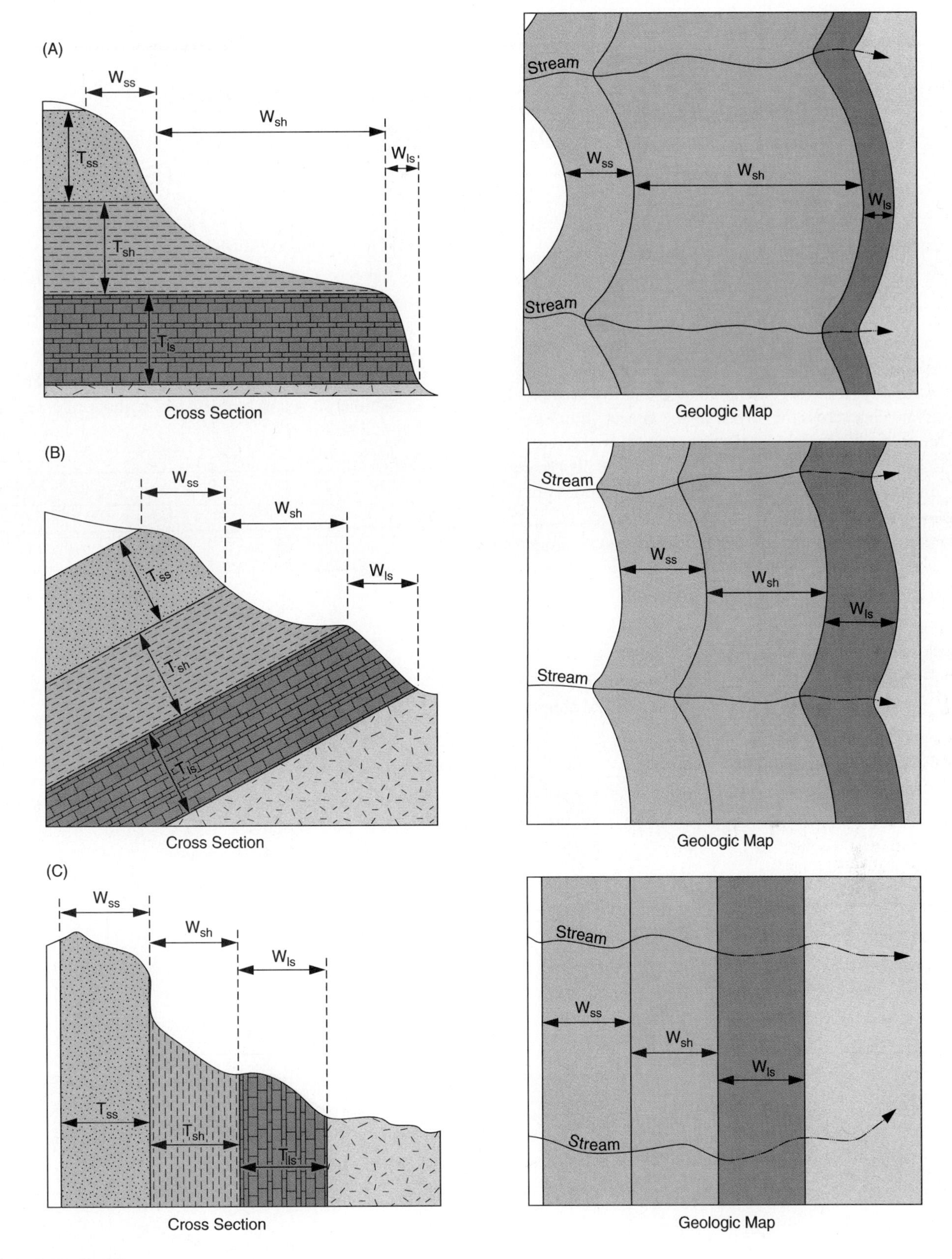

FIGURE 4.10

Cross section and geologic maps of three sedimentary formations showing the relationship of the true thickness of each formation (T_{ss}, T_{sh}, and T_{ls}) to their widths of outcrop on a geologic map. The thickness of each formation is the same on all three cross sections. The width of outcrop (W_{ss}, W_{sh}, and W_{ls}) on the geologic map depends on the thickness and attitude of the formations and the slope of the surface. In the three cases shown—(*A*) horizontal strata, (*B*) inclined strata, and (*C*) vertical strata—only in C does the thickness of each formation in the cross section equal the width of outcrop of the same formation on the geologic map.

Name Section Date

EXERCISE 20

GEOLOGIC STRUCTURES ON BLOCK DIAGRAMS, GEOLOGIC SYMBOLS, AND RELATIVE AGES OF FORMATIONS

1. Complete the four block diagrams in figure 4.11. Below each block diagram, print the name of one or more of the geologic structures shown. Remember that the numbers on the map indicate the relative ages of the formations, with number 1 being the oldest. Assume that the topography in all four diagrams is essentially flat except in diagrams *C* and *D* where a stream cuts across the formational contacts. In these four block diagrams, dip directions in (A) and (B) are to be determined by the rule that older beds dip toward younger beds, and in (*C*) and (*D*), dip directions are to be determined by the rule of V's.
2. On figure 4.7, place a strike and dip symbol at points *a, b, c, d,* and *e* on the geologic map and the block diagram. Use a black pencil.
3. Figure 4.8 shows two geologic maps. All formations shown there are sedimentary in origin.
   (a) Label each formation with a number indicating its relative age in the sequence of strata. The oldest should be labeled number 1.
   (b) Draw strike and dip symbols on each map.
4. On the geologic maps of figure 4.10*A* and *B*, draw the appropriate geologic symbol that shows the attitude of each of the three formations and show by numbers the relative age of each.
5. Why is it impossible to tell the relative ages of the formations shown in figure 4.10*C* without reference to figure 4.10*A* or *B*?

____

____

____

____

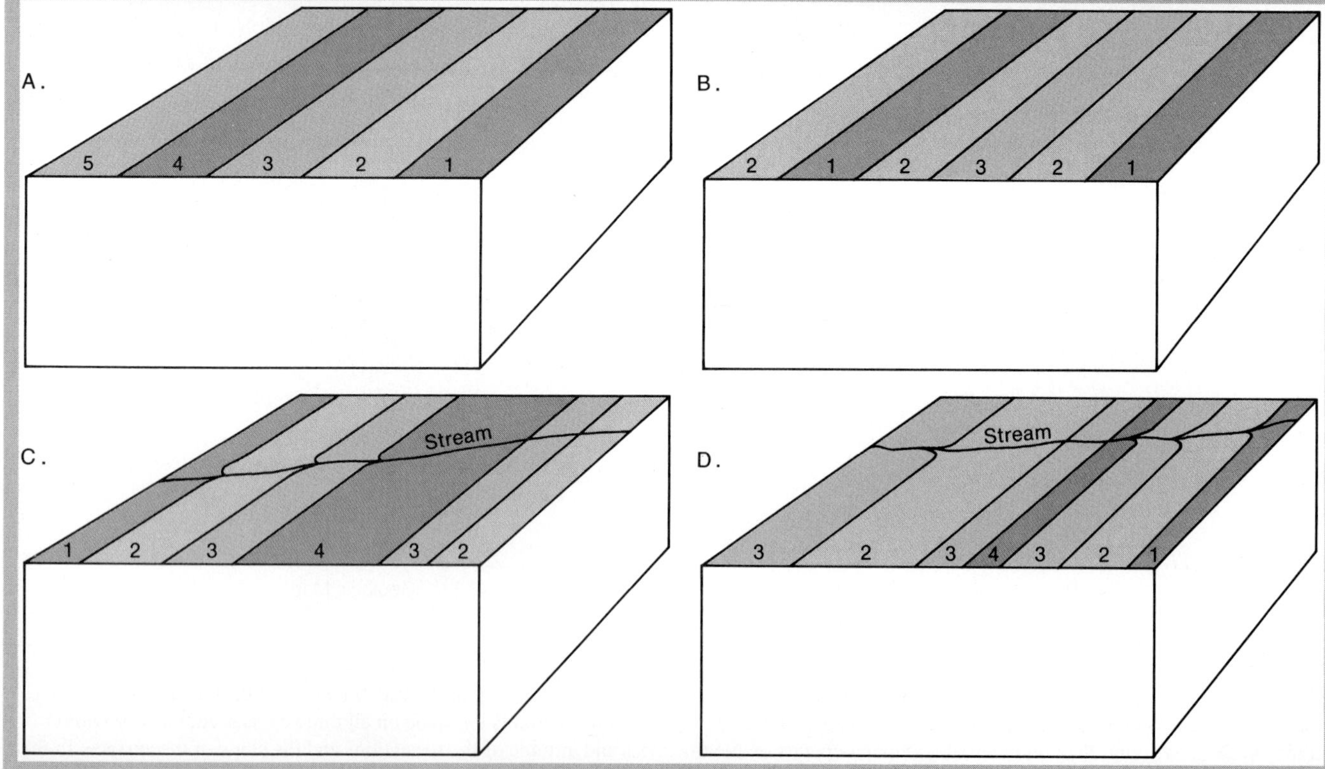

FIGURE 4.11

Block diagrams for use in Exercise 20, Question 1.

Name Section Date

EXERCISE 21A

GEOLOGIC MAPPING ON AERIAL PHOTOGRAPHS

CHASE COUNTY, KANSAS

Figure 4.12 shows the outcrop pattern of sedimentary strata. The light and dark gray tones correspond to different lithologic characteristics of the various formations. At *A*, near the center of the photograph, the contacts of a light gray formation are shown by two black lines.

1. Extend these lines along the contacts as far as possible on the photo. Do the same for formation *B* shown in the upper left-hand corner of the photo.
2. What is the general attitude of these two formations? (Compare the outcrop pattern of fig. 4.12 with fig. 4.9*A*.)

____

____

____

3. If it is assumed that the thickness of these two formations is constant, why do their widths of outcrop change from place to place?

____

____

____

4. Applying the law of superposition in this case, which of the two formations is relatively older than the other?

____

____

____

FIGURE 4.12

Aerial photograph, Chase County, Kansas. Scale, 1:20,000.

(U.S. Dept. of Agriculture photograph.)

Name Section Date

EXERCISE 21B

GEOLOGIC MAPPING ON AERIAL PHOTOGRAPHS

FREMONT COUNTY, WYOMING

This stereopair (fig. 4.13) shows sedimentary strata cropping out in the area. Study the stereopair with a stereoscope, and using figure 4.9B for guidance, answer the following questions.

1. Draw several strike and dip symbols on the right-hand photo of the stereopair and write a verbal description of the attitude of the formations.

___

___

___

___

2. Are the oldest beds in the northern or southern part of the area? What rule is applied here that allows you to answer this question?

___

___

___

___

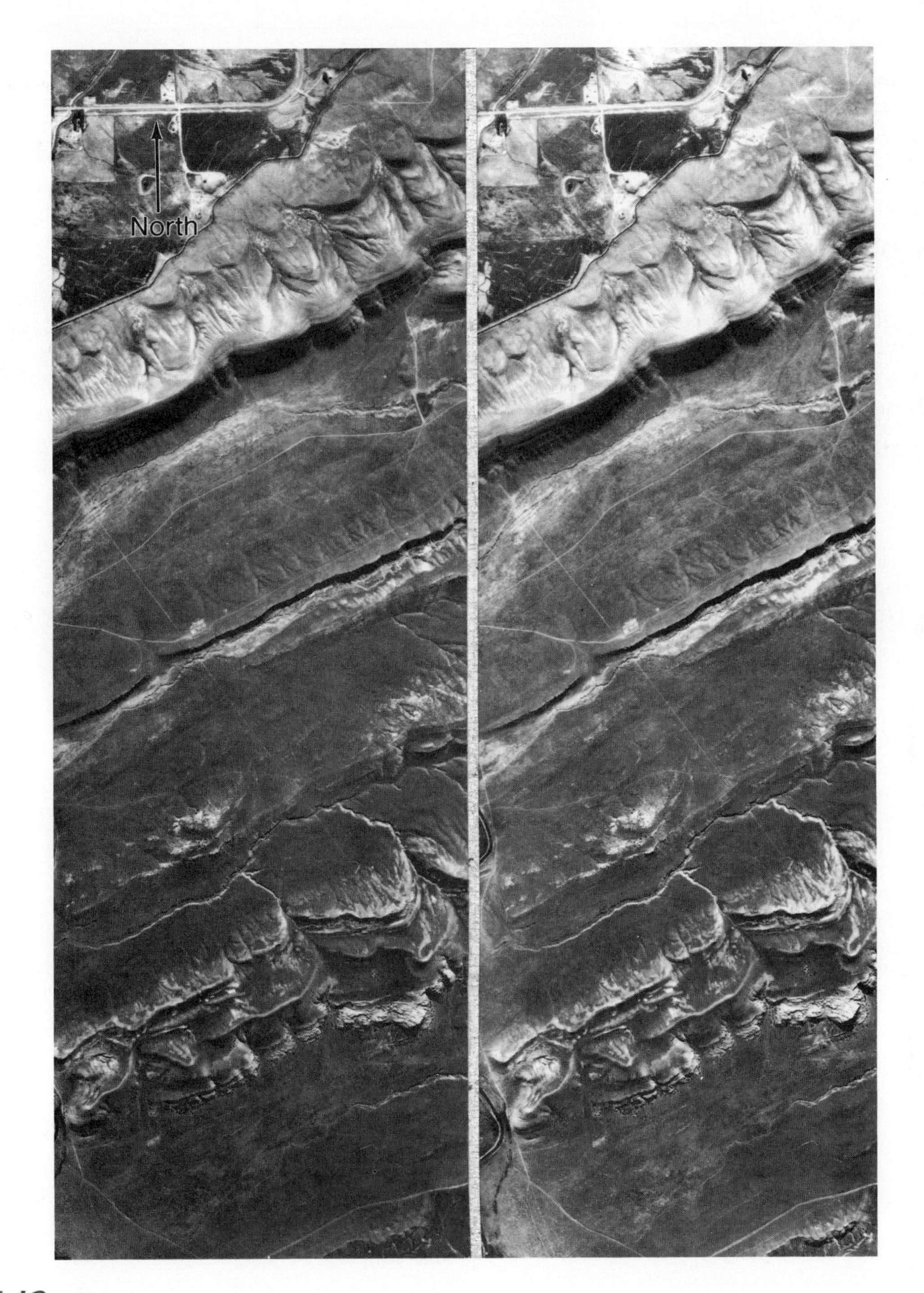

FIGURE 4.13

Stereopair of part of Fremont County, Wyoming. Scale, 1:21,500; July 13, 1960.
(U.S.G.S.)

Name Section Date

EXERCISE 21C

Geologic Mapping on Aerial Photographs

Aerial Photograph, Arkansas

The pattern of curved ridges in this photograph (fig. 4.14) is the result of differential erosion of sedimentary strata. The ridges are composed of rocks that are more resistant to erosion than the rocks that form the intervening valleys. The structure displayed in the photograph is the nose of a steeply plunging anticline.

1. Draw the trace of the axial plane of the fold on the photograph with a red pencil and add other appropriate symbols on the axial trace and elsewhere on the photograph to indicate all relevant structural information. (Refer to fig. 4.2 as a reminder of the appropriate symbols to use for the fold axis.)

FIGURE 4.14

Aerial photograph showing the nose of a plunging anticline in Arkansas. Scale, 1:24,000; November 9, 1957.
(U.S.G.S.)

Name Section Date

EXERCISE 21D
GEOLOGIC MAPPING ON AERIAL PHOTOGRAPHS

LITTLE DOME, WYOMING

The structure shown here (fig. 4.15) is an elongate dome or a doubly plunging anticline. Use a stereoscope to study the stereopair while formulating the answers to the following questions.

1. Draw strike and dip symbols on the right-hand photograph of the stereopair. Use red pencil.
2. Draw the trace of the axial plane and other symbols that are appropriate for this structure on the right-hand photograph. Use red pencil.
3. What is the evidence that the angle of dip changes as one follows the ridges northward along the eastern flank of the structure?

____
____
____
____

4. If a hole were drilled on the axis of the fold at the center of the structure, would the drill encounter any of the formations that crop out on the surface in the area covered by the photographs? Explain.

____
____
____
____

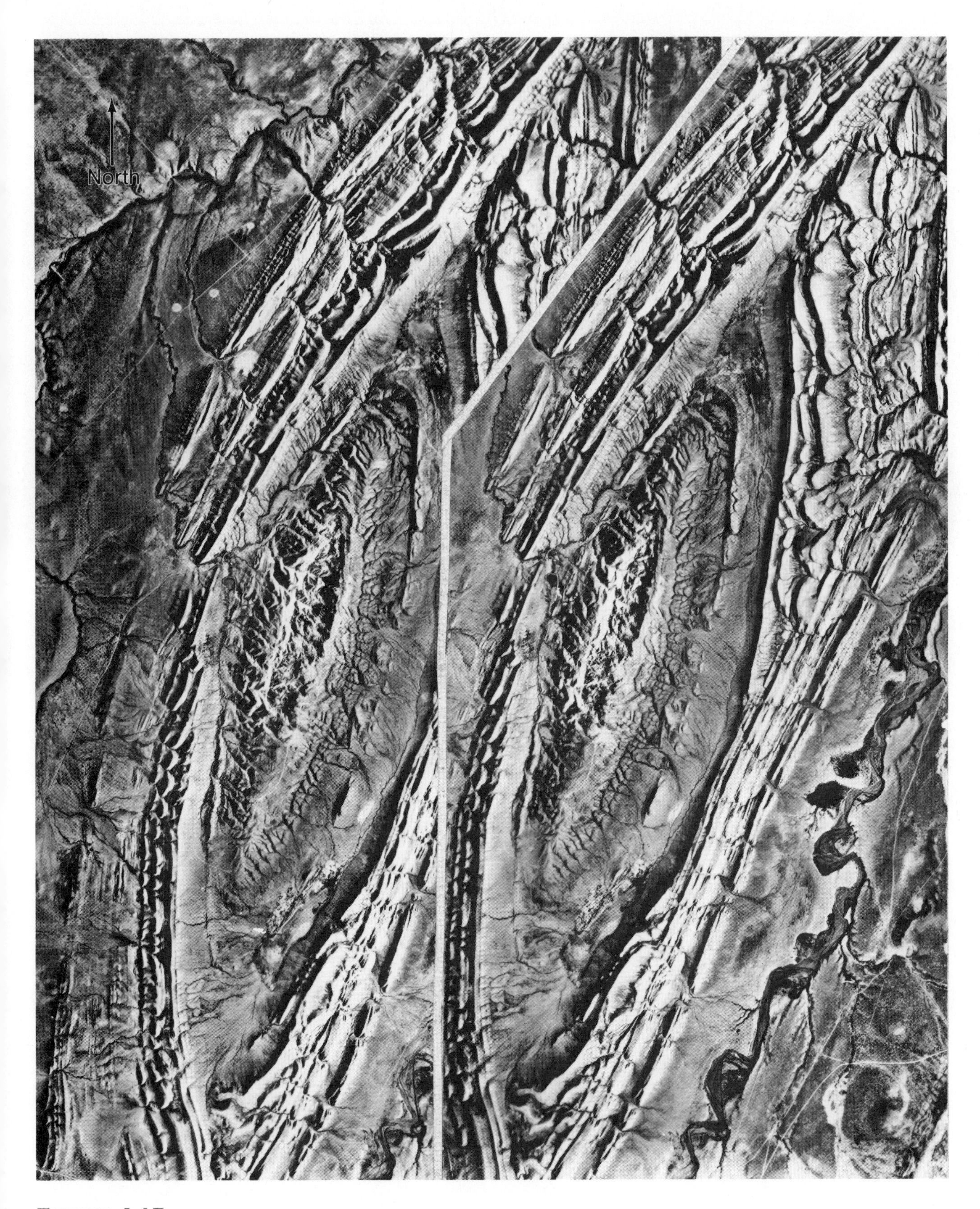

FIGURE 4.15

Stereopair of Little Dome, Wyoming. Scale, 1:23,600; October 20, 1948.
(U.S.G.S.)

Name Section Date

EXERCISE 21E
GEOLOGIC MAPPING ON AERIAL PHOTOGRAPHS

HARRISBURG, PENNSYLVANIA

Figure 4.16 looks like an aerial photograph but, in fact, is a side-looking airborne radar (SLAR) image that enhances surface features. A SLAR image is particularly useful in deciphering the structural geology of an area.

The patterns of ridges and valleys shown here are the result of differential erosion of a series of plunging folds in the Appalachian Mountains of Pennsylvania. The river flowing through the area is the Susquehanna River, and the city of Harrisburg lies at the upper margin of the figure near the river.

Figure 4.16 is oriented so that the north direction is toward the lower margin of the figure. The reason for this is that the "shadows" produced by the imaging process must be toward the viewer in order for the ridges on the ground to appear as ridges on the image. (By turning the page upside down, you may see a "reverse topography," that is, the ridges appear as valleys.) Examine the image to get a feel for the structural pattern it reveals.

1. Locate the zig-zag ridge that is cut by the Susquehanna River at four places on the image. Using an easily erasable pencil, draw the trend of this formation on the figure. Or, to put it another way, draw a continuous strike line along the crest of the ridge throughout its length. For the purpose of identification, this ridge-forming formation will be called formation *A*. It forms the flanks of plunging synclines where it is intersected by the Susquehanna River.
2. When you are satisfied that you have identified formation *A* by the method just described, draw over your pencil line with a yellow felt-tipped pen or yellow pencil to distinguish the ridge from other ridges in the figure.
3. Locate the *next youngest* ridge-forming formation and trace its strike as you did for formation *A*. This younger formation will be called formation *B*, and it should be identified with a color on the image that contrasts with the one used for formation *A*.
4. Using a pencil, draw the axial traces of the folds that occur on the figure showing the direction of plunge and the symbol for an anticline or syncline. Reinforce your pencil line with a red pencil when you are certain of your interpretation.
5. Are the rocks west of the Susquehanna River (i.e., between the river and the right-hand margin of the figure) generally older or younger than those near the left-hand margin of the figure? Explain your reasoning.

____

____

____

____

____

FIGURE 4.16

Side-looking airborne radar (SLAR) image mosaic of part of the Harrisburg map, Pennsylvania. Scale, 1:250,000.

(U.S. Geologic Survey. Synthetic-Aperture Radar Imagery. Experimental Edition, 1982.)

EXERCISE 22A

INTERPRETATION OF GEOLOGIC MAPS

LANCASTER GEOLOGIC MAP, WISCONSIN

Six formations are shown on this map (fig. 4.17), each of which is identified by an abbreviation. The abbreviations and the names they represent are, in alphabetical order: Od, Decorah Formation; Ogl, Galena Dolomite; Op, Platteville Formation; Opc, Prairie du Chien Group; Osp, St. Peter Sandstone; and Qal, Alluvium. A *group* consists of two or more formations with significant features in common.

The contacts of these formations are more or less parallel to the topographic contours, thereby indicating that the formations are more or less horizontal. Another set of contours, shown in red, defines the top of the Platteville Formation (Op). The red numbers associated with these red contour lines indicate the elevation of the contour line above sea level.

1. Determine the oldest and youngest formations on the map and those of intermediate age. (Label oldest with number 1.) Complete the geologic column in figure 4.17 by printing the *abbreviation* of a formation in the appropriate box and printing the *name* of the formation on the line immediately below the box.
2. Using the topographic contour lines, estimate the thickness of the Decorah Formation (Od), the Platteville Formation (Op), and the St. Peter Sandstone (Osp). Print the estimated thickness in feet of each of these formations to the right of the appropriate box in the geologic column. The thickness of a formation is determined by subtracting the elevation of the bottom of the formation from the elevation of the top of the formation. These elevations can be estimated from contour lines on either side of a contact.
3. Why is it impossible to determine the thickness of the Galena Dolomite and the Prairie du Chien Group?

____

____

____

____

4. Locate Cement School in the northeast part of the map area. A road intersection near the school has an elevation of 1,076 feet and is so marked on the map. Using a nearby contour line showing the top of the Platteville Formation, determine how deep a well must be drilled at the road intersection to reach the top of the Platteville Formation.

____

____

____

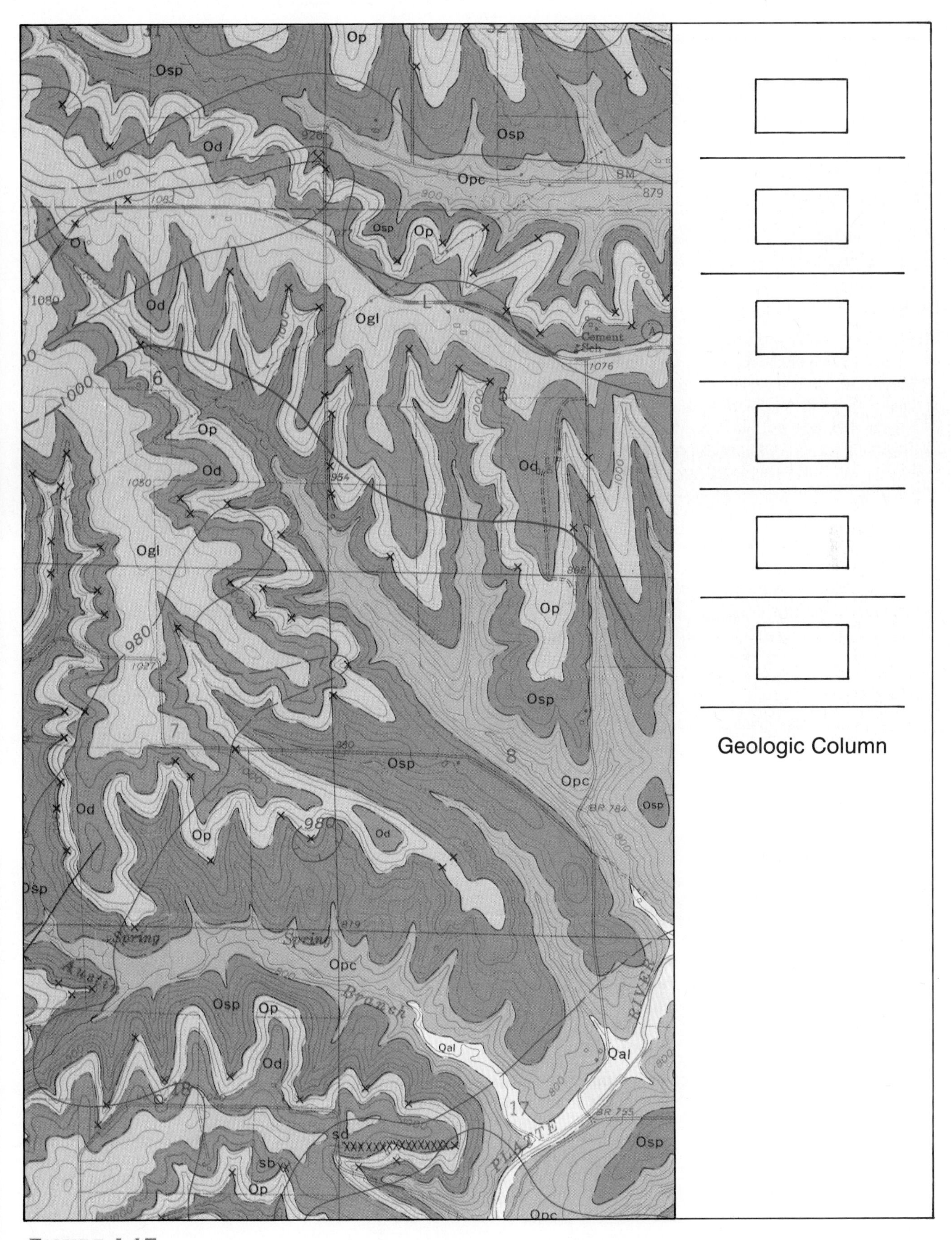

FIGURE 4.17

Geologic map of part of the Lancaster quadrangle, Grant County, Wisconsin, 1971. Scale, 1:24,000; contour interval, 20 feet. U.S. Geological Survey.

Name Section Date

EXERCISE 22B
INTERPRETATION OF GEOLOGIC MAPS

SWAN ISLAND GEOLOGIC MAP, TENNESSEE

This geologic map (fig. 4.19) shows a banded outcrop pattern of sedimentary rocks. Study the map to get a general feel for the geology. Notice the strike and dip symbols on the map.*

1. The topographic profile of figure 4.18 is drawn along the black line on the map from the northwest to the southeast. Draw a geologic cross section along this line using figure 4.18 as the base. In aligning the profile of figure 4.18 with the line of profile on the map, place the point on the profile labeled Briar Fork at the point on the map where the creek (one inch from the lower right hand corner of the map) crosses the line of profile.

   The formations on the map are identified with letter symbols. To simplify the drawing of the geologic cross section, we will treat some of the formations as a group and show them as a group rather than as individual formations on the cross section. We will establish four arbitrary groups, each labeled by a letter symbol, and retain two single formations as they appear on the map. The groups and formations to be used in the cross section are as follows:

   Group Є formations Єm, Єn, Єmn, and Єcr.
   Group LO: formations Oc, Ok, and Oma.
   Group MO: formations A, B, C, D, EFG, H, I, J, K, and LM.
   Formation Omb.
   Formation Os.
   Group SM: formations Sc and Mdc.

   On the cross section, label each group of formation according to the above scheme. Ignore formation Qal.

2. Complete the geologic column at the right of the map by placing the symbols for the four groups and two formations shown on your cross section in the six boxes provided. Use standard geologic practice—the youngest at the top and the oldest at the bottom.

The number that appears next to a strike and dip symbol on a geologic map refers to the angle of dip of the rock layers as measured by a field geologist at a specific rock outcrop or exposure. When these numbers occur near a line on the map along which a geologic cross section is to be constructed, they should be considered approximations of the dip angles rather than absolute values. Dip angles within a few tens of feet of each other can vary as much as 5 to 10 degrees.

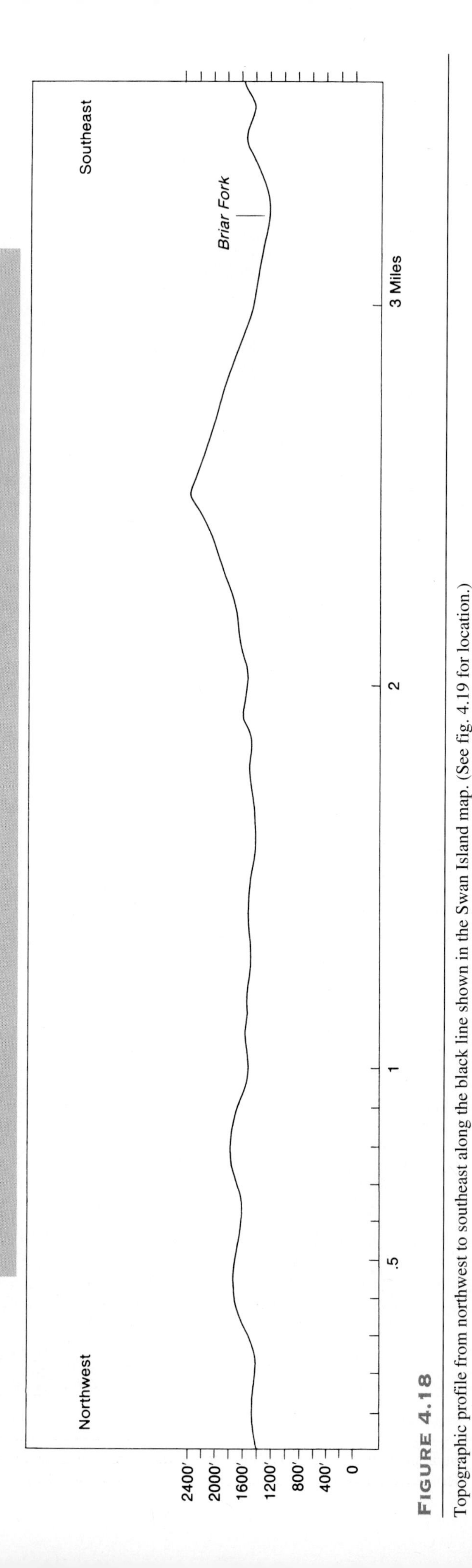

FIGURE 4.18

Topographic profile from northwest to southeast along the black line shown in the Swan Island map. (See fig. 4.19 for location.)

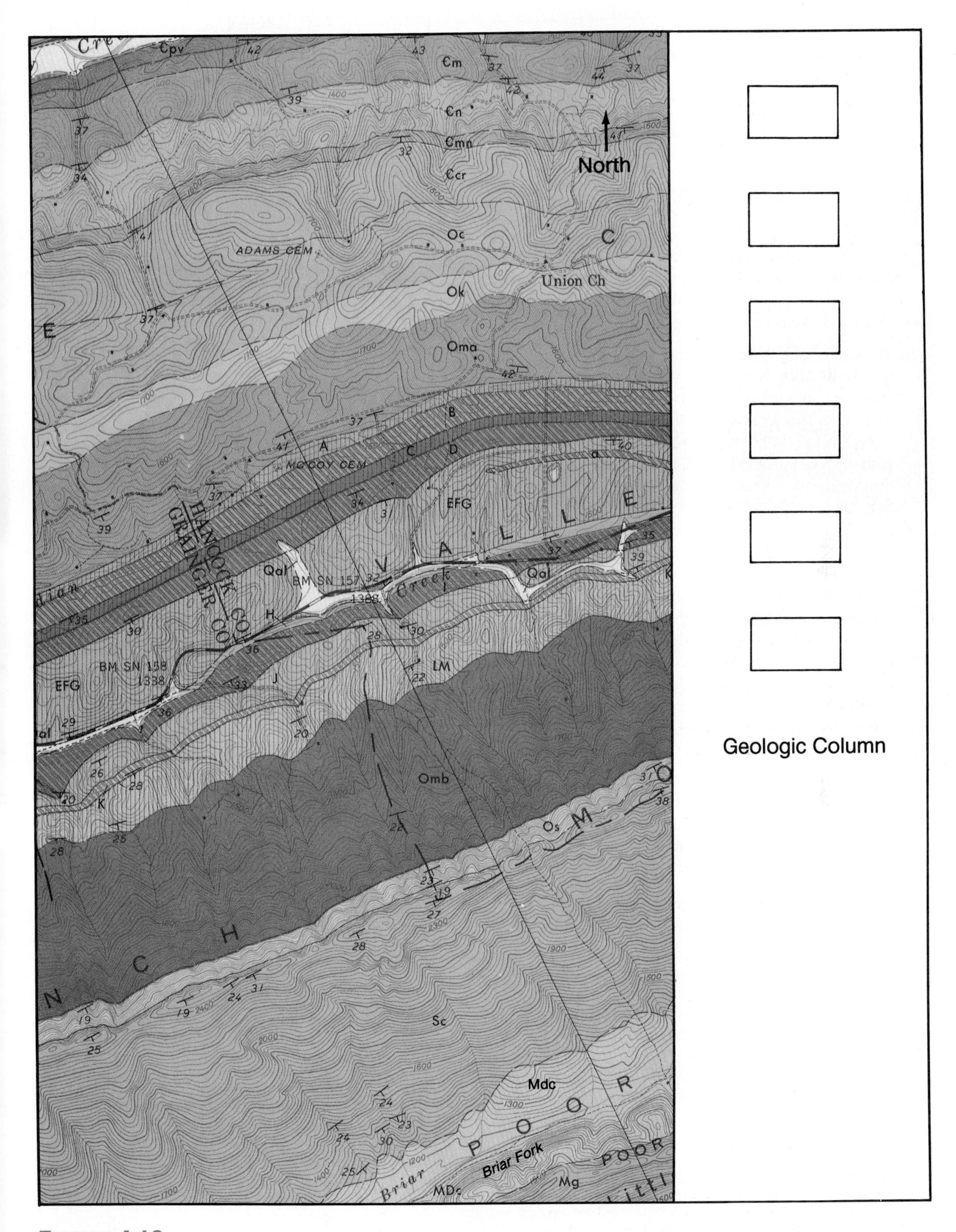

FIGURE 4.19

Geologic map of part of the Swan Island quadrangle, Tennessee, 1971. U.S. Geologic Survey. Scale, 1:24,000; contour interval, 20 feet. The black line is the line of topographic profile of Figure 4.18.

Name Section Date

EXERCISE 22C
INTERPRETATION OF GEOLOGIC MAPS

COLEMAN GAP GEOLOGIC MAP, TENNESSEE-VIRGINIA

This area (fig. 4.21) is underlain by sedimentary rock formations of different thicknesses. Note the many strike and dip symbols that occur on the map. (North is toward the top of the page.)

1. The topographic profile of figure 4.20 is drawn along the line trending northwest to southeast across the map area from margin to margin, passing through the location of Brooks Well. Draw a geologic cross section along this line. Align the Brooks Well on the profile with the position of the Brooks Well on the map to achieve the proper correlation between the topography of the profile and the contours on the map. Note that the vertical scale of the profile is identical with the horizontal scale of the map. For assistance in drawing the geologic cross section, it should be noted that the Brooks Well penetrated the base of the €c formation 300 feet below the ground surface. Color formations €c and €cr on the cross section, and label all formations with their correct symbols.
2. What is the thickness of formation €cr?

    ____

    ____

3. Complete the geologic column to the right of the map. Put the symbol of the formation in the appropriate box and the name of the formation on the line below the box. The formational names and their respective symbols are as follows:
    €c, Conasauga Shale
    €cr, Copper Ridge Dolomite
    €mn, Maynardville Limestone
    Ocl, Lower Chepultepec Dolomite

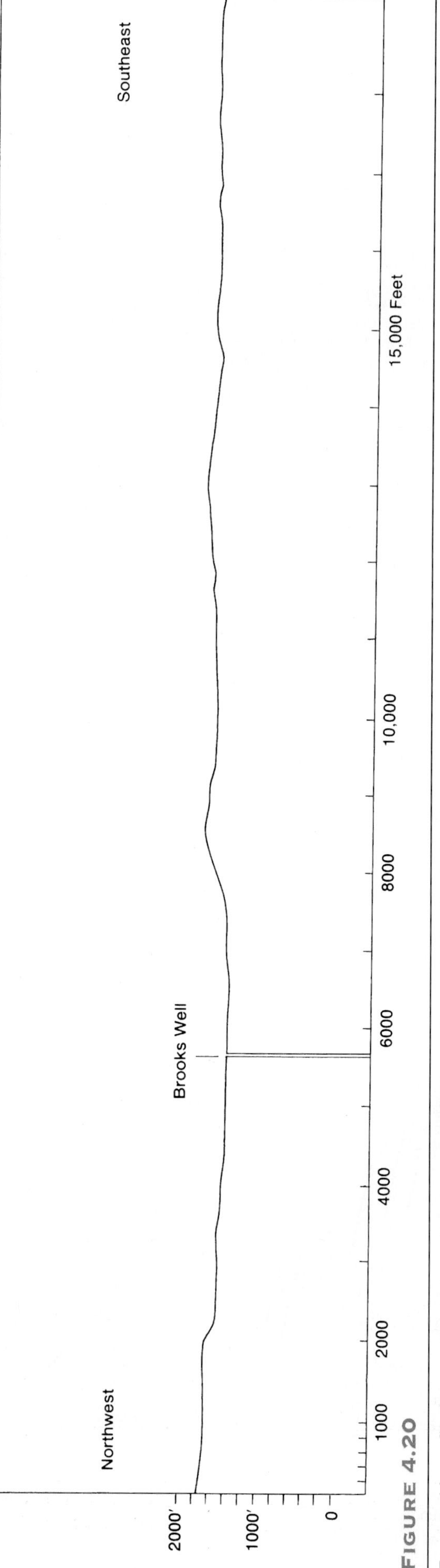

FIGURE 4.20
Topographic profile from northwest to southeast across the Coleman Gap map. (See fig. 4.21 for location.)

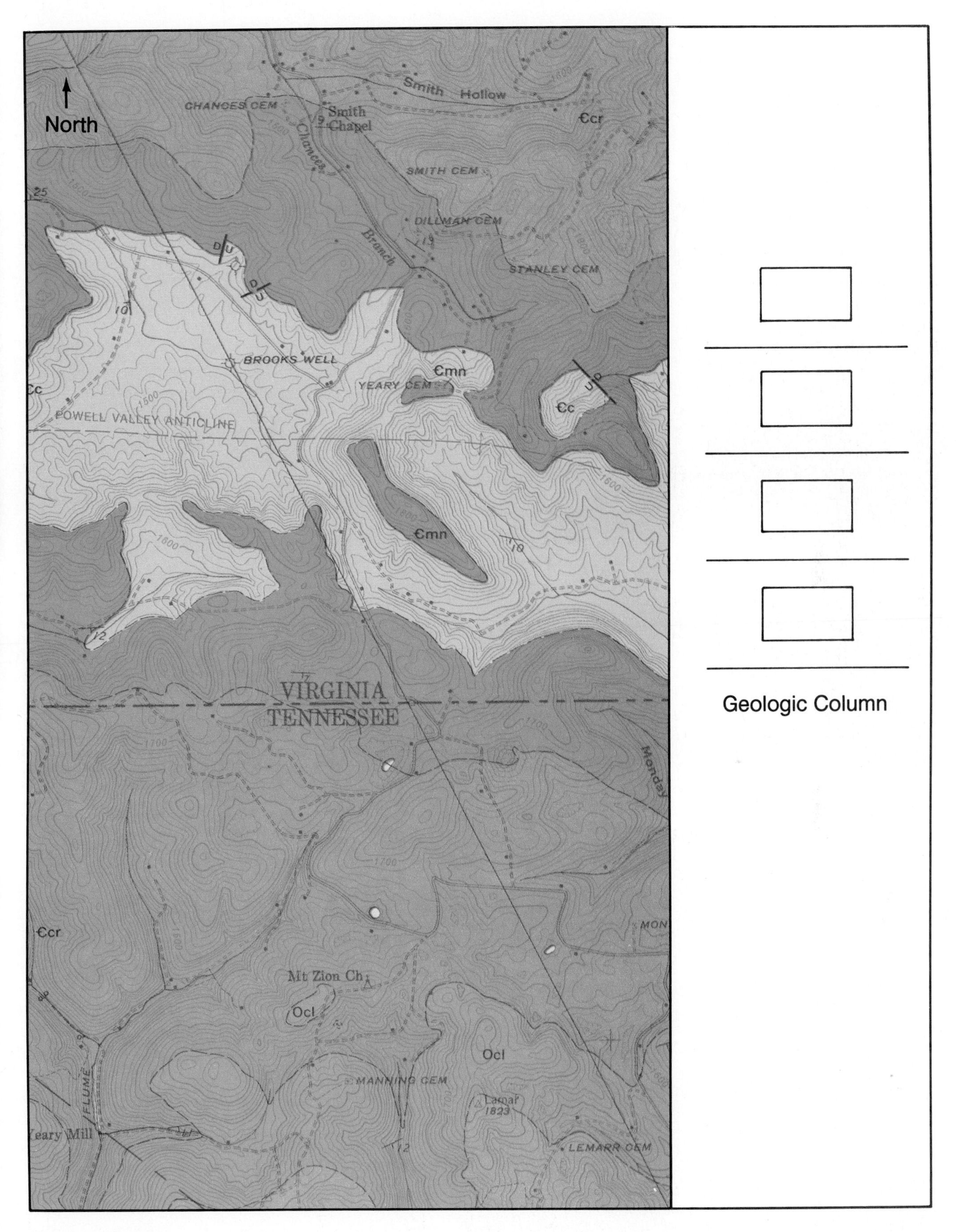

FIGURE 4.21

Geologic map of part of the Coleman Gap quadrangle, Tennessee-Virginia, 1962. U.S. Geologic Survey. Scale, 1:24,000; contour interval, 20 feet. (The black line passing through the Brooks Well profile is the line of topographic profile of figure 4.20.)

Faults and Earthquakes

Earth stresses that produce folds also produce faults. A *fault* is a fracture or break in the earth's crust along which differential movement of the rock masses has occurred. Movement along a fault causes dislocation of the rock masses on each side of the fault so that the contacts between formations are terminated abruptly.

Faults may be active or inactive. *Active faults* are those along which movement has occurred sporadically during historical time. Earthquakes are caused by movement along active faults. *Inactive faults* are those in which no movement has occurred during historical time. They are treated as part of the structural fabric of the earth's crust.

In this section we will deal first with inactive faults as part of structural geology and, secondly, with active faults and their relationship to earthquakes.

Inactive Faults

A fault is a planar feature, and therefore its attitude can be described in much the same way that any geologic planar feature can be described. Figure 4.2 shows the various symbols used on a geologic map to define faults.

Nomenclature of Faults

Figure 4.22 is a block diagram of a hypothetical faulted segment of the earth's crust. The *fault plane* is defined as *abcd.* The fault plane strikes north-south and dips steeply to the east. A single horizontal sedimentary bed acts as a reference marker and shows that the displacement along the fault plane is equal to the distance *x–y*. This is called the *net slip.* The arrows show the direction of relative movement along the fault plane. Block *A* has moved up with respect to block *B*, and conversely, block *B* has moved down with respect to block *A*. Block *A* is called the *upthrown side* of the fault, and block *B* is the *downthrown side.* Block *B* is also known as the *hanging wall,* and block *A* as the *footwall.* Both terms are derived from miners who drove tunnels along fault planes to mine ore that had been emplaced there.

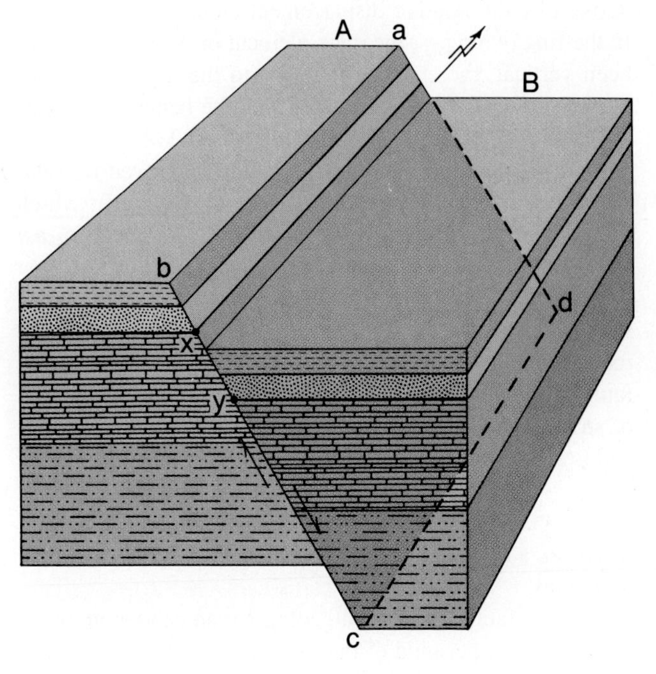

FIGURE 4.22

Block diagram of a fault. Arrows show the relative movement of block *A* with respect to block *B*. The horizontal beds have been dislocated a distance of *x–y*. The fault plane is *abcd.*

Faults generally disrupt the continuity or sequence of sedimentary strata, and they cause the dislocation of other rock units from their prefaulted positions. On geologic maps, the intersection of the fault plane with the ground surface is called a *fault trace.* Fault traces are depicted on geologic maps by the use of standard symbols (fig. 4.2).

After faulting occurs, erosion usually destroys the surface evidence of the fault plane, so that with the passage of time the *fault scarp* (the exposed surface of the fault plane in figure 4.22) is destroyed. Only the fault trace remains.

Types of Faults

Faults are divided into three major types, each of which is defined by the relative displacement along the fault plane. In the first two types, the main element of displacement has been vertical, more or less parallel to the dip of the fault plane. A *normal fault* is one in which the hanging wall has moved down relative to the footwall (fig. 4.23*A*). A *reverse fault* is one in which the hanging wall has moved up relative to the footwall (fig. 4.23*B*). A reverse fault in which the fault plane dips less than 45 degrees is called a *thrust fault.*

The third category of faults is characterized by relative displacement along the fault plane in a horizontal direction parallel to the strike of the fault plane. This type of fault is called a *strike-slip fault* (fig. 4.23*C*). A special type of strike slip fault, a *transform fault,* is described on p. 242 and is shown in figure 5.1.

Figure 4.23*D* shows a *horst,* an upthrown block bounded on its sides by normal faults. Figure 4.23*E* shows a *graben,* a downthrown block bounded on its sides by normal faults.

A fault shown on a geologic map can be analyzed to determine what kind of fault is involved. The analysis of normal and reverse faults will reveal the hanging and footwalls that lead to an understanding of the relative movement along the fault plane. In a strike-slip fault, off-setting of a marker bed as in figure 4.23*C* is the most direct evidence of the direction of movement. In this case the displacement is to the right as one looks across the fault and this would be called a *right-lateral fault.* If movement was to the left as you look across the fault, it would be a *left-lateral fault.*

A normal or reverse fault that cuts across the strike of inclined or folded sedimentary beds presents one of the most common situations for the analysis of movement along the fault plane. In such cases, there will be an apparent migration of the beds in the direction of dip of these beds on the upthrown side of the fault as erosion progresses. Stated another way, if an observer were to stand astride the fault trace, the observer's foot resting on the older rock would rest on the upthrown side. This is a simple mental test that can be applied to the analyses of faults presented in Exercise 23.

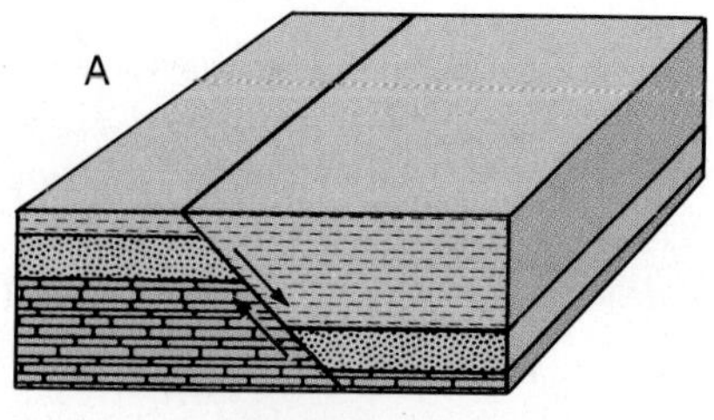

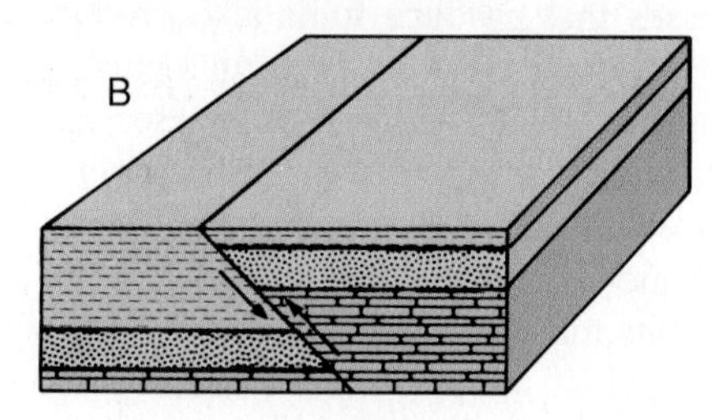

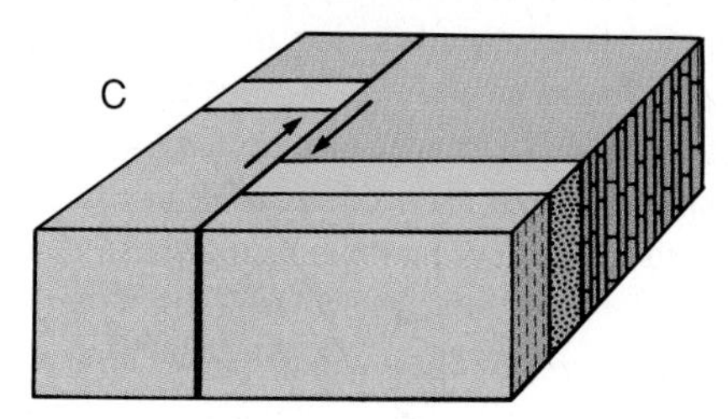

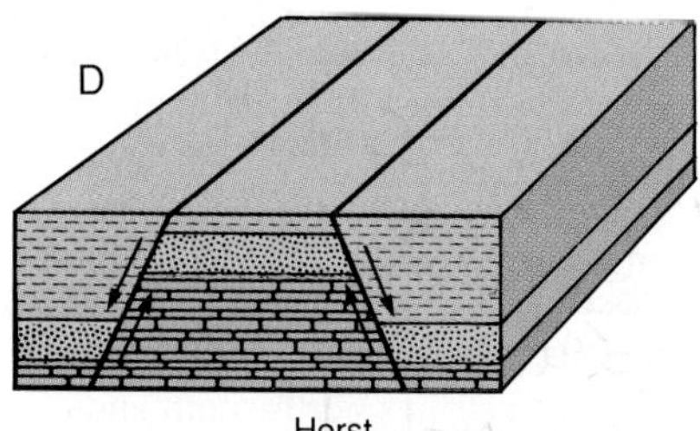

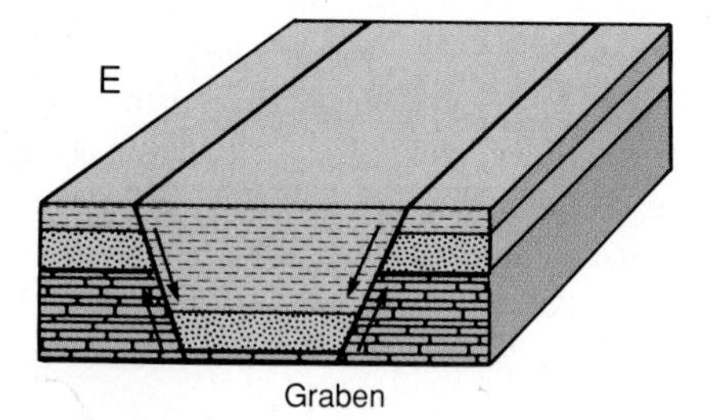

Figure 4.23

Block diagrams illustrating major fault types. Arrows indicate relative movement along the fault plane. Note that these diagrams illustrate the map and cross section views ***after*** erosion has removed the fault scarp so that on the surface only the fault trace remains.

EXERCISE 23B
FAULT PROBLEMS

FAULTED SEDIMENTARY STRATA

1. On the Swan Island map (fig. 4.27), formation Ok is cut by a fault trending NW-SE. Label the upthrown and downthrown sides of the fault with the correct geologic symbols.
2. Refer to the fault described under question 1. What would be the direction of *dip* of the fault plane if this were a normal fault?

   ____

   ____

3. If the fault referred to in question 1 were a reverse fault, what would be the direction of dip of the fault?

   ____

   ____

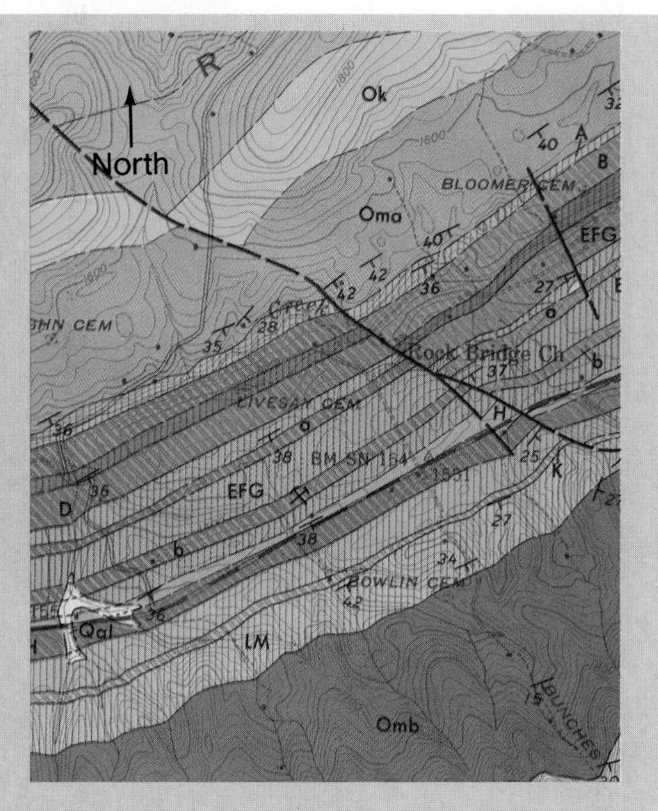

FIGURE 4.27

Part of the geologic map of Swan Island quadrangle, Tennessee, 1971. U.S. Geological Survey. Scale, 1:24,000; contour interval, 20 feet.

THE USE OF SEISMIC WAVES TO LOCATE THE EPICENTER OF AN EARTHQUAKE

SEISMOGRAPHS, SEISMOGRAMS, AND SEISMIC OBSERVATORIES

The energy released by an earthquake produces vibrations in the form of *seismic waves* that are propagated in all directions from the focus. Seismic waves can be detected by an instrument called a *seismograph,* and the record produced by a seismograph is called a *seismogram.* The geographical location of a seismograph is called a *seismic station* or *seismic observatory,* and it is given a name in the form of a code consisting of three or four capital letters that are an abbreviation of the full name of the station. For example, a seismic observatory on Mt. Palomar, in southern California, has the code designation of PLM. A worldwide network of seismic observatories provides records of the times of arrival of the various kinds of seismic waves. Seismograms from at least three different stations located around the focus of a given earthquake but at some distance from it provide the data needed to locate the epicenter.

SEISMIC WAVES

Two general types of seismic waves are generated by an earthquake: *body waves* and *surface waves.* Body waves travel from the focus in all directions through the earth; they penetrate the "body" of the earth. Surface waves travel along the surface of the earth and do not figure in the location of an epicenter.

Body waves consist of *primary waves* and *secondary waves.* The primary wave is referred to as the *P wave,* and the secondary wave is the *S wave.* The P wave is like a sound wave in that it vibrates in a direction parallel to its direction of propagation. An S wave, on the other hand, vibrates at right angles to the direction of wave propagation (fig. 4.33).

P and S waves are generated at the same time at the focus, but they travel at different speeds. The P wave travels almost twice as fast as the S wave and is always the first wave to arrive at the seismic station. The S wave follows some seconds or minutes after the first arrival of the P wave. *The difference in arrival times of the P and S waves is a function of the distance from the seismic station to the epicenter.* The distance from the seismic station to the epicenter is called the *epicentral distance.*

READING A SEISMOGRAM

A seismograph records the incoming seismic waves as wiggly lines on a piece of paper wrapped around a drum rotating at a fixed rate of speed. The resulting seismogram contains not only the record of the incoming seismic waves but also marks that indicate each minute of time. Clocks at all seismic stations around the world are set at Greenwich Mean Time (GMT), so no matter what time zones observatories are located in, the seismograms produced at them are all based on a standardized clock.

When no seismic waves are arriving at an observatory, the seismograph draws a more or less straight line (fig. 4.34). Some small irregular wiggles on the seismogram may be *background noise* from vibrations produced by trucks, trains, heavy surf, construction equipment, and the like. Most modern seismographs contain a damping mechanism that reduces background noise to a minimum. In addition, background noise is kept to a minimum if the observatory is located in a remote area where human activities are uncommon.

The time of arrival of the first P wave is noted as T_P. The P wave continues to arrive until the first S wave appears, which is noted as T_S. The S wave has a much larger amplitude than the P wave. (The amplitude is the vertical

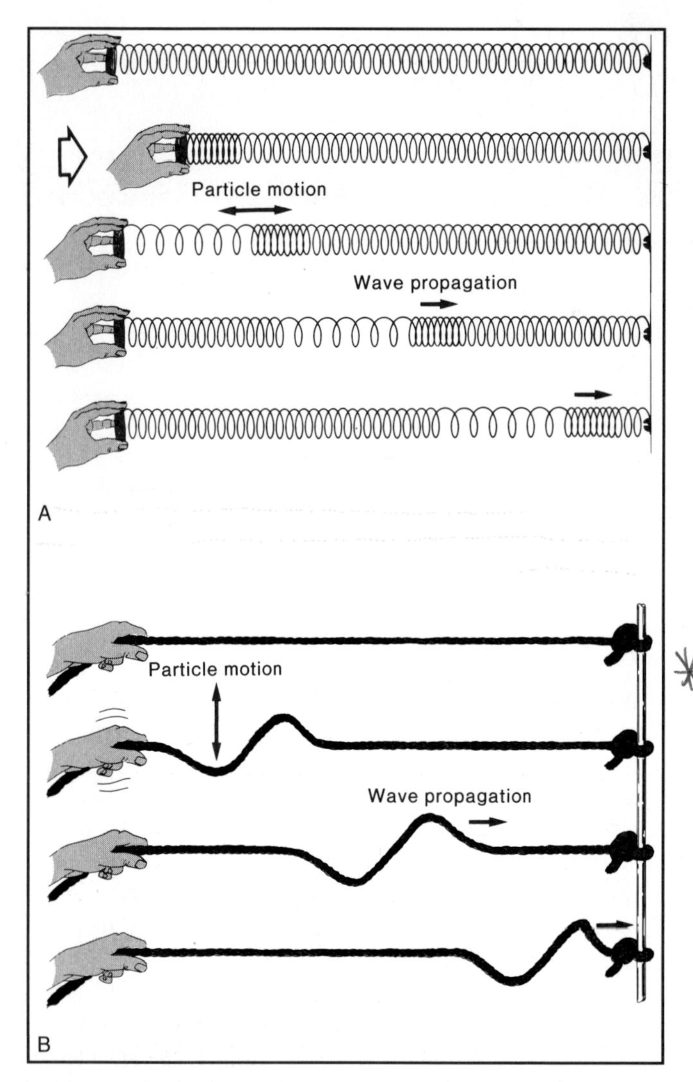

FIGURE 4.33

Particle motion in seismic waves. (*A*) P wave is illustrated by a sudden push on the end of a stretched spring. The particles vibrate *parallel* to the direction of wave propagation. (*B*) S wave is illustrated by shaking a loop along a stretched rope. The particles vibrate *perpendicular* to the direct wave propagation.

(From Charles C. Plummer and David McGeary, Physical Geology, 6th ed.

distance between the peak of the recorded wave and the line on the seismogram recorded when no seismic waves are arriving.)

Figure 4.34 shows a seismogram from the Santa Ynez Peak Observatory on which the arrival times of the P and S waves are shown as T_P and T_S. These were determined by using the time scale on the seismogram to measure the time from the mark labeled 12:40:00 (12 hrs: 40 min: 00 sec) to the times of arrival of the first P and S waves. On the SYP seismogram of figure 4.34, T_P is 19 seconds after the time mark, or 12:40:19 GMT, and T_S is 44 seconds after the time mark, or 12:40:44 GMT. The difference between the time of arrivals, $T_S - T_P$, is therefore 25 seconds.

Locating an Epicenter on a Map Using Travel-Time Curves

$T_S - T_P$ is measured in units of time, and this time, when converted to a distance, indicates the epicentral distance. Converting this time to distance requires the use of *travel-time curves* for both the P and S waves as shown in figure 4.35. A point on either one of the curves indicates the time required for a P or S wave to travel a certain distance from the epicenter. Time in seconds is shown on the vertical scale, and the corresponding distance in kilometers is shown on the horizontal scale. Following is the procedure for converting $T_S - T_P$ in seconds to an epicentral distance in kilometers:

1. Determine T_P and T_S from a seismogram to the nearest second. Record these values for use in the next step.
2. Subtract T_P from T_S and record as a time in seconds for use in the next step.
3. From the vertical scale of the travel-time graph of Figure 4.35, determine the length of a line with the value of $T_S - T_P$. Mark this distance on a sheet of paper.
4. Keep the paper vertical and move the paper upward and to the right until one end of the line lies on the P curve and the other end of the line lies on the S curve. Follow the vertical line down to the horizontal scale and read the epicentral distance.

   As an example of this procedure, let us use the data from the SYP seismogram of figure 4.34. The value for $T_S - T_P$ on this seismogram is 25 seconds. The length of the line corresponding to 25 seconds is marked on the paper from the vertical travel time axis and the paper is moved to the right until one end is on the P curve and the other end is directly above it on the S curve. Follow this vertical line down and read 192 kilometers, the epicentral distance at station SYP.
5. On a suitable base map use the bar scale to set your compass to the epicentral distance determined in step 4. Use this compass setting to draw a circle on the map whose center is at the geographic coordinates of the appropriate seismic station.
6. By following steps 1 through 5 for three different seismograms at appropriate directions and distances from the epicenter, you will draw three circles that intersect or nearly intersect at the epicenter.

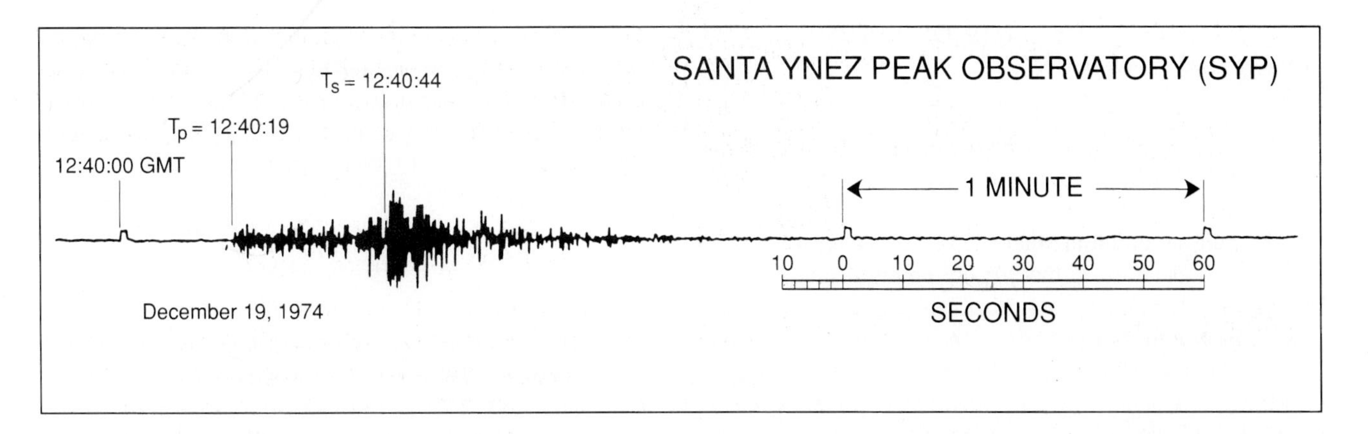

FIGURE 4.34

A seismogram recorded at the Santa Ynez Peak Observatory (SYP) in California showing an earthquake on December 19, 1974. The time mark automatically recorded on the seismogram is 12:40:00, which is 12:40 P.M. Greenwich Mean Time (GMT). The time of arrival of the first P wave, T_P, is 12:40:19, and the time of arrival of the first S wave, T_S, is 12:40:44.

(Source: Based on data from Charles G. Sammis, University of Southern California.)

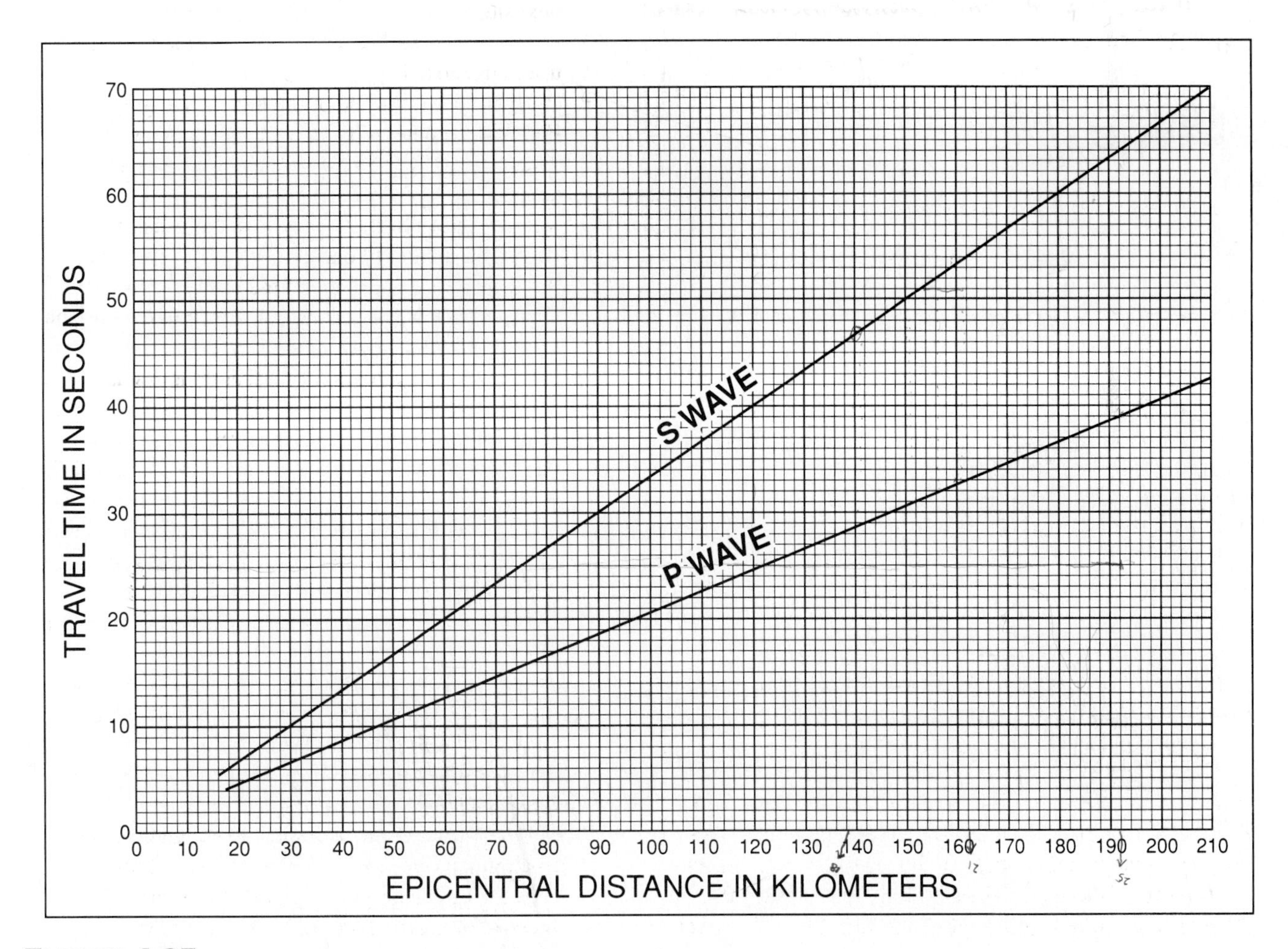

FIGURE 4.35

Travel-time curves for P and S waves in southern California.

(Source: Based on data from Charles G. Sammis, University of Southern California.)

Determining the Time of Origin of an Earthquake

The epicentral distances determined from $T_s - T_p$ are used to find the time of origin of the earthquake, designated by the symbol T_o. The procedure to do this is best explained by an example. Let us return to the information from the seismogram, recorded at station SYP, of figure 4.34. We have already determined that the epicentral distance is 192 kilometers. Remember that this is the distance from the seismic station to the epicenter of the earthquake. We want to know the time when this earthquake occurred, T_o. That is also the time when the seismic waves started their journey of 192 kilometers to SYP. Looking at figure 4.35, we see that the point where the P wave curve intersects the 192-kilometer line is 39 seconds. This tells us that it took the P wave 39 seconds to travel from the earthquake epicenter to station SYP. T_o is determined by subtracting the travel time of the P wave, 39 seconds, from T_p which is 12:40:19 GMT. Subtracting 39 seconds from 12 hrs, 40 minutes, 19 seconds gives us 12:39:40 GMT, the time of origin of the earthquake, or T_o.

References

Bolt, Bruce. 1978. *Earthquakes: A primer.* W. H. Freeman & Company, San Francisco, Chapter 6.

Eiby, George A. 1980. *Earthquakes.* Van Nostrand Reinhold Company, New York. 209 pp.

We are indebted to Charles G. Sammis, Department of Geological Sciences at the University of Southern California, for his assistance in preparing this exercise.

PART FIVE

PLATE TECTONICS AND RELATED GEOLOGIC PHENOMENA

BACKGROUND

The theory of plate tectonics is a widely accepted concept that has been guiding geophysical and geological research since the mid-1960s. The term *plate tectonics* refers to the rigid plates that make up the skin of the earth and their movement with respect to one another. Figure 5.1 shows the distribution of the principal plates as they are currently understood. The plates differ greatly in size, although the true size of the plates is distorted in figure 5.1 because the map on which they are displayed is a Mercator projection. This kind of map represents the spherical shape of the earth on a flat surface and shows extreme distortion in the polar regions because the longitude lines on the map are parallel instead of converging (see fig. 2.1). It can be seen from figure 5.1 that some plates include both oceans and continents, as for example the Africa and South America plates.

Plate tectonics explains many phenomena on planet Earth, both on the continents and in the ocean basins, such as the origin and distribution of volcanoes and earthquakes, the topography of the sea-floor, and a host of other major geologic features.

THE MAJOR COMPONENTS OF THE EARTH

Through the use of geophysical techniques it has been determined that the earth is composed of layers or shells of different composition and mechanical properties. The compositional layers of the earth are the *crust, mantle,* and *core* (figure 5.2). The crust is subdivided into *continental crust,* which is mainly granitic in composition, and *oceanic crust,* which is mainly basaltic. The mechanical layers are the *lithosphere,* composed of the crust and upper mantle; the *asthenosphere,* a weak, ductile (a ductile substance is one that deforms without fracturing) layer of the earth's upper mantle (beneath the lithosphere) on which the lithospheric plates move; the *mesosphere,* the lower mantle; the "liquid" metallic *outer core;* and the solid metallic *inner core.*

The relationship of these features is shown in figure 5.2, a scale diagram of a cross section of the earth from the surface to the core. The information shown on figure 5.2 has been gleaned mainly from the interpretation of seismic waves and laboratory experiments on rocks under high temperatures and pressures. None of the features in figure 5.2 have been observed directly in place, except, of course, the outermost crust at the earth's surface. Figure 5.2 is therefore an interpretation of the conditions prevailing at depth beneath the surface of the earth, based on the current theory of plate tectonics.

PLATE MOVEMENT

The arrows in figure 5.2 show the direction of movement of the four plates covered in the cross section. This movement is deduced from the characteristics of the plate boundaries. Two plates that are moving away from each other form a *divergent* or *spreading boundary* along an *oceanic ridge* (Mid-Atlantic Ridge) or *rise.* Basaltic magma is extruded at these boundaries to form new oceanic crust. Two plates that are moving toward each other form a *subduction zone* (Peru-Chile Trench), where oceanic lithosphere descends beneath continental lithosphere along a deep ocean trench. Another kind of plate boundary is one where two plates slide past each other, as in the case of the San Andreas fault, which forms the boundary between the Pacific and the North America plates (fig. 5.1).

The forces that cause plate movement are not clearly understood, but the consensus is that *convection currents* are formed by differential heating of the mantle, which causes the asthenosphere to act like a giant conveyor belt along which the plates are rafted.

Earth scientists believe that the surface area of the earth has remained constant over geologic time. Therefore, as new crustal material is formed along

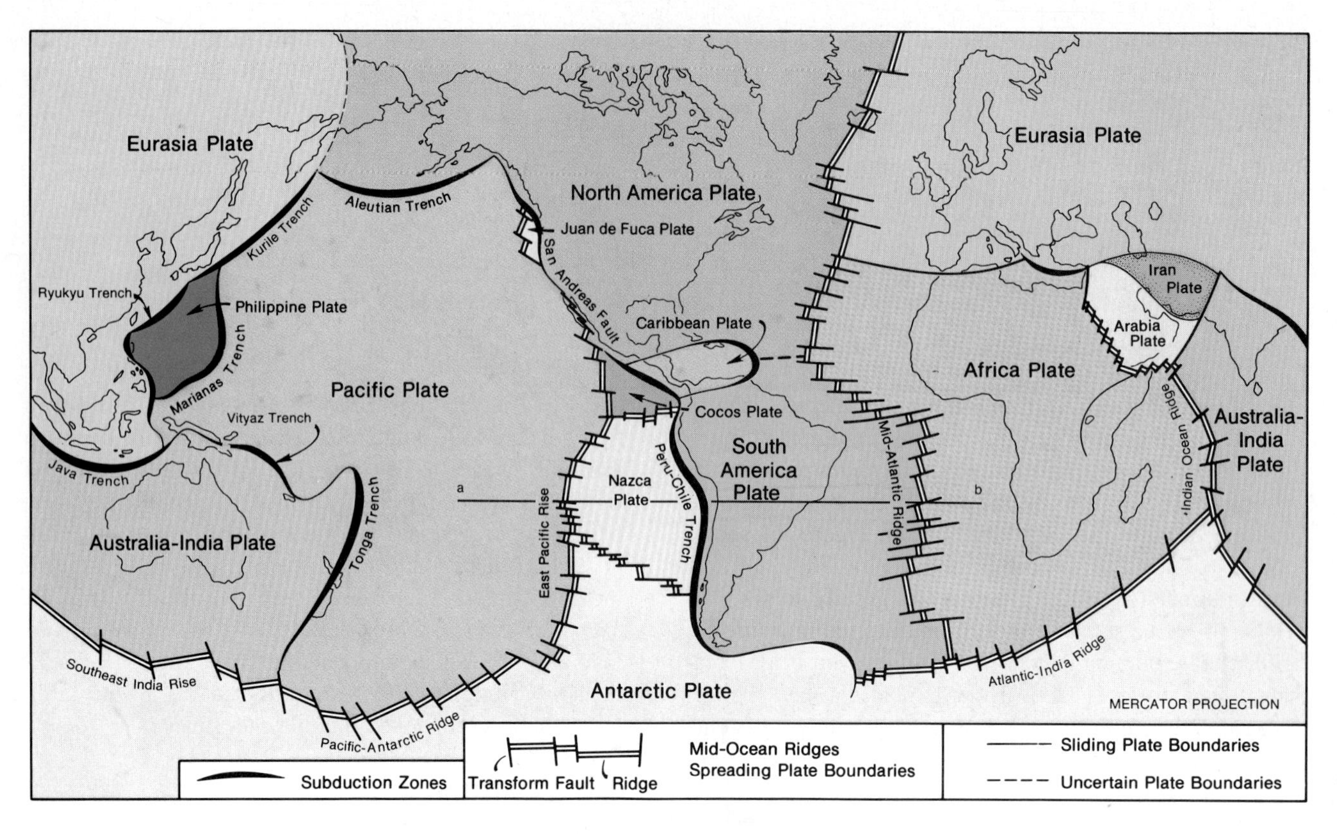

FIGURE 5.1

Map of the earth showing the names and boundaries of major lithospheric plates. The line labeled a–b is the line of the cross section of figure 5.2.

divergent boundaries by the extrusion of basalt from the underlying mantle, the creation of this new crust must be compensated by destruction of crust elsewhere. This is in fact what appears to happen in subduction zones where oceanic lithosphere slowly descends beneath adjacent continental lithosphere, as shown in figure 5.2 at the boundary between the Nazca and South America plates. The descending cool and rigid lithospheric slab is gradually incorporated into the hot mantle.

A larger scale schematic drawing of a hypothetical subduction zone is shown in figure 5.3. The oceanic lithosphere of basaltic crust and rigid upper mantle is subducted beneath the continental lithosphere. The foci of earthquakes in the subducted plate indicate that the subducted slab is still rigid; that is, it fractures or faults instead of flowing as is the case with the ductile asthenosphere. The earthquake foci in figure 5.3 are generalized to show that the shallow earthquakes are concentrated near the top of the descending plate and that the deeper earthquakes occur lower in the slab. This suggests that the upper part of the descending plate loses its rigidity as it moves deeper into the asthenosphere. The foci of earthquakes do not occur below about 700 kilometers. This fact supports the idea that the subducted slab loses its rigidity by the time it has reached that depth; it has literally been consumed in the mantle.

Earthquakes associated with divergent plate margins have shallow foci that are concentrated along the actively spreading ridge and the *transform faults* (see page 220 in Part 4) that offset the ridges. The pattern of these shallow earthquakes therefore roughly defines the location of the divergent plate margins.

Magma is generated above the subducted plate as the wet oceanic crust melts in contact with the hot asthenosphere. It then rises toward the surface where it forms plutons or may reach the surface to form volcanoes, most commonly of andesitic composition.

Seafloor Spreading in the South Atlantic and Eastern Pacific Oceans

Background

A basic premise of plate tectonics is that the crustal plates have moved with respect to each other over geologic time and, in fact, are moving today. The rates of movement of crustal plates can be determined by using data from the plate margins along the mid-ocean ridges, where the amount of movement can be measured.

To measure the movement of two adjacent crustal plates along the margins of a divergent plate boundary, two things must be known: (1) two points on adjacent diverging plates that were once at the same geographic coordinates but have since moved away from each other over a known distance, and (2) the time required for the two points to move from their original coincident position to their present positions. If the two points can be identified and plotted on a map, the distance between them can be measured by use of the map scale. Determining the age in actual years of the two points involves the earth's magnetic field. It is therefore necessary to review this subject in order to understand how it relates to the movement of crustal plates.

The Earth's Magnetic Field

The earth is encompassed by a magnetic field. The source of this magnetic field is in the liquid metal outer core. The field generated is analogous to the lines of force produced by a bar magnet with a north pole at one end and a south pole at the other. Imagine an immense bar magnet passing through the center of the earth with its north and south poles located near the North and South poles of the earth's axis of rotation (fig. 5.7*A*). A magnetic compass placed in this field would align itself parallel to the lines of magnetic force. The direction of this force is shown by the arrows in figure 5.7*B*, which point from the south magnetic pole toward the north magnetic pole. This condition is called *normal polarity.*

A magnetic compass does not point to the north geographic pole (true north) however, because the magnetic poles are not coincident with the geographic poles. The geographic poles define the earth's axis of rotation and remain fixed with respect to the equator over geologic time. The magnetic poles, however, shift over time with respect to the geographic poles. A plot of the magnetic poles during historical time shows that they tend to stay in close proximity to the geographic poles, so on a geological time scale, it is assumed that the magnetic poles and the geographic poles have remained within about 10 degrees of each other.

Reversals of the Magnetic Poles

Lava flows may contain minerals with magnetic properties that align themselves in the earth's magnetic field while the lava is still molten. When the lava solidifies, the magnetite minerals remain aligned parallel to the lines of force in the earth's magnetic field that existed when the lava cooled into rock. These magnetic minerals are like minute magnetic compasses frozen in the rock. Other rocks, like sandstone, also contain magnetic minerals that become aligned with the existing magnetic field when they sink to the lake bottom or seafloor at their site of deposition.

One of the amazing features of the earth's magnetic field is that its polarity has reversed itself many times over geologic time. That is, the north and south magnetic poles abruptly changed places so that a magnetic compass would point to the south magnetic pole during periods of *reversed polarity* (fig. 5.7*C*).

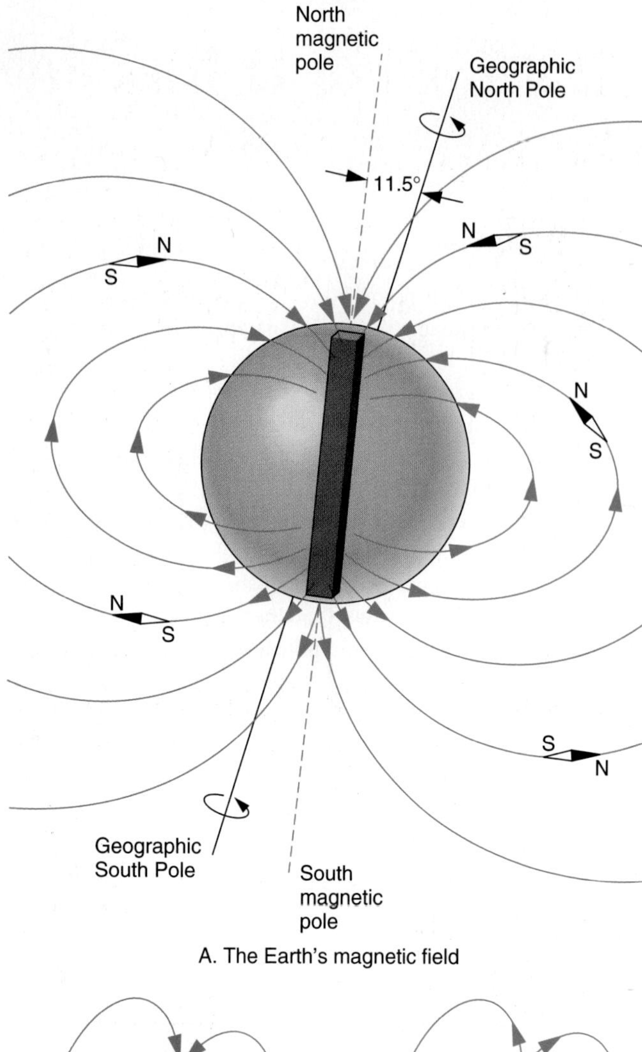

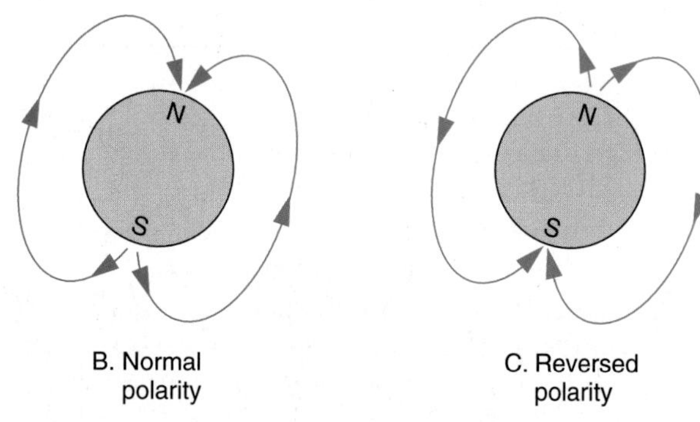

FIGURE 5.7

The earth's magnetic field. (*A*) The north magnetic pole and the north geographic pole are not coincident, and over geologic time they have been not much further apart than they are today. (*B*) Normal polarity characterizes the earth's magnetic field when the direction of the magnetic lines of force is from the south magnetic pole toward the north magnetic pole. (*C*) Reversed polarity occurs when the direction of the magnetic lines of force is toward the south magnetic pole. The earth's magnetic field has reversed many times during the history of the earth.

(...harles C. Plummer and David McGeary, Physical Geology, 6th ed. ...1993 McGraw-Hill Company, Inc., Dubuque, Iowa. All rights reserved. ...ion.)

The study of magnetism in ancient rocks is called *paleomagnetism.* The paleomagnetic features of rocks studied over the entire world have provided the basis for a detailed chronology of times of normal and reversed polarities during the last 170 million years. The rock layers from which this chronology has been assembled have been dated by radioactive means, thereby providing an absolute time scale that identifies the times when the periods of normal and reversed polarities occurred.

Both normal and reversed polarities are called *magnetic anomalies* or simply *anomalies.* A chronology of magnetic anomalies is given in figure 5.8. Three elements are contained in this chronology: (1) a time scale in millions of years before the present (Ma Age in figure 5.8), (2) the periods of normal (black) and reversed (white) polarities, and (3) the conventional identification numbers that have been assigned to each anomaly. These identification numbers are arranged in chronological order with the youngest anomaly designated by number 1 and successively older anomalies by numbers 2, 2a, 3, and on up to anomaly 33. (Figure 5.8 is a shortened version of a chronology that extends to anomaly 37, which is about 170 million years old.)

The time scale of figure 5.8 will be used later in an exercise, so it will be useful to become familiar with it. For example, the anomaly with the identification number 6 is a positive anomaly that is 20 million years old. Notice that the identification number assigned to an anomaly bears no relationship to that anomaly's absolute age. In other words, the series of identification numbers represents a *relative chronology,* and the series of dates in millions of years is an *absolute chronology.*

Magnetic Anomalies

An anomaly is a departure from the normal scheme of things. With respect to rocks that contain magnetic minerals, a *magnetic anomaly* is a magnetic reading that is greater or less than the normal strength of the magnetic field where the rock occurs. To illustrate, consider a series of bands of lava beds (basalt) lying on the seafloor along both sides of an active spreading ridge between two crustal plates. An instrument that measures the intensity of the magnetic field is called a *magnetometer.* When one is towed on a long cable behind a ship headed along a course across the trend of the lava beds, the strength of the earth's magnetic field is recorded continuously on board the ship, and a satellite navigational system simultaneously records the ship's position.

When the seaborne magnetometer passes over basalts that are normally polarized, the strength of the earth's magnetic field is slightly intensified because the ancient magnetism in the normally polarized rocks adds a small component to the normally polarized earth's field. In

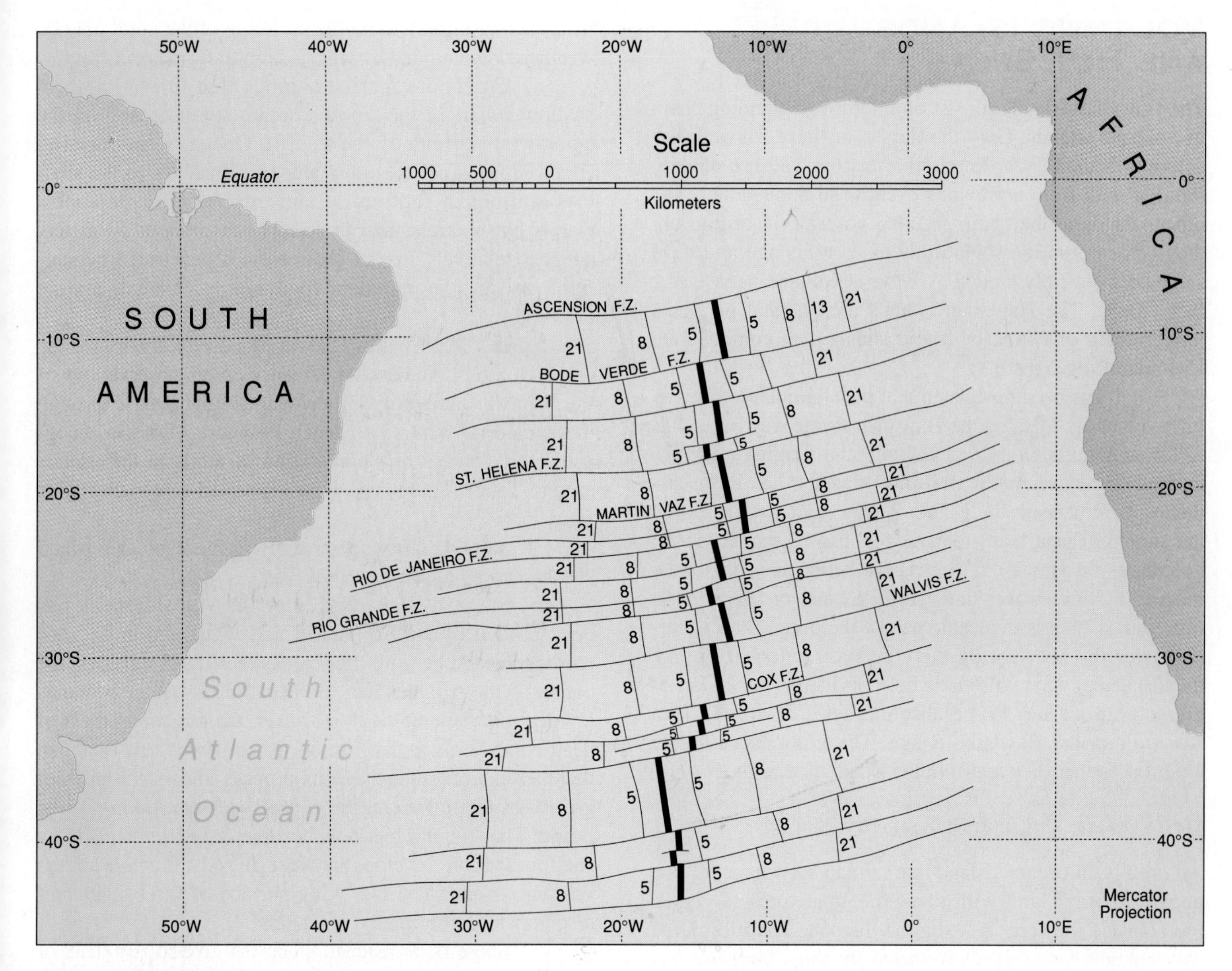

FIGURE 5.12

Map of the South Atlantic Ocean showing part of the Mid-Atlantic Ridge in black bars, the east-west fracture zones, and selected magnetic anomalies. The ages of the numbered anomalies or magnetic lineations can be determined from the magnetic time scale in figure 5.8.

(From Magnetic Lineations of the World's Ocean Basins. Copyright © 1985 the American Association of Petroleum Geologists, Post Office Box 979, Tulsa, Oklahoma 74101. Reprinted by permission.)

WEB CONNECTIONS

http://www.ldeo.columbia.edu/~menke/plates.html

Enter longitude and latitude from a point near one of the major ridges and this site will calculate a spreading rate for you.

http://sideshow.jpl.nasa.gov/mbh/series.html

Use of the Global Positioning System (GPS) to determine horizontal velocities mostly due to motion of the earth's tectonic plates. Images with vectors for California and the globe.

Volcanic Islands and Hot Spots

The Hawaiian Islands consist of a northwest-trending chain of volcanic islands. Only the largest of these, Hawaii, has active volcanoes, whose eruptive history you are already familiar with from a previous exercise in this manual. The other islands in the chain are also volcanic in origin, but they ceased erupting some millions of years ago and have since been severely eroded by wave action and intense surface runoff. The Hawaiian Islands are, in fact, part of a longer chain of extinct volcanic islands that compose the Hawaiian Ridge (fig. 5.4).

It has been postulated that the alignment of the extinct volcanoes forming the Hawaiian Ridge was formed as follows. A *hot spot,* whose latitude and longitude has remained fixed over many millions of years, lies in the asthenosphere beneath the island of Hawaii. This hot spot is the source of heat that produces the magma that feeds the volcanoes on Hawaii. The oceanic lithosphere has been moving northwest over this spot in a more or less straight line, and as each nonvolcanic part of the ocean floor is carried over the hot spot by the conveyor action of the asthenosphere, a new volcano is born. Evidence in support of this hypothesis lies in the absolute dates of the old lava flows along the Hawaiian Ridge. These are successively older the farther they are from the island of Hawaii.

Volcanic Islands and Atolls

An *atoll* is an oceanic island that in map view appears as a narrow strip of land with low relief that forms a closed loop. Inside the loop is a shallow lagoon. The loop itself may contain gaps that allow access by ships from the surrounding ocean to the lagoon. An atoll is made chiefly of *coral reef,* but it also contains assorted other marine organisms such as algae and mollusks. Some of the coral may be weathered and eroded by wave action to form *coral sand.*

Corals are marine animals that thrive in warm shallow waters of the world's oceans, most notably in the equatorial regions of the Pacific Ocean. Reef-forming corals grow in colonies that attach themselves to the shallow seafloor in subtropical and tropical climatic zones. Corals live in water that is no more than about 50 meters deep, is relatively free of sediments, is penetrated by sunlight, and has an abundant food supply of small marine organisms.

In 1842, Charles Darwin proposed a theory for the origin of atolls. He based this theory on his observations of the tropical islands during the voyage of the *Beagle* through the equatorial waters of French Polynesia. Darwin recognized three stages in the evolution of atolls in the islands around Tahiti. These stages are illustrated schematically in figure 5.13.

Stage I consists of a newly formed volcanic island surrounded by a *fringing coral reef.*

Stage II is reached after the volcanic peak has been eroded by surface runoff and wave action and has partly subsided beneath the sea. As the island subsides, the corals of the fringing reef die because the water becomes too deep for their survival. However, the remaining mass of dead coral forms a platform on which new corals establish themselves continuously. This process allows the upward growth of the reef to maintain pace with the sinking of the island. The fringing reef now becomes a *barrier reef,* and a shallow lagoon develops between it and the shore of the volcanic island. The low-lying surface of the barrier reef also may be colonized by vegetation.

Stage III is reached when the eroded remnants of the volcanic peak disappear beneath the sea through continued surface erosion and subsidence of the seafloor. The

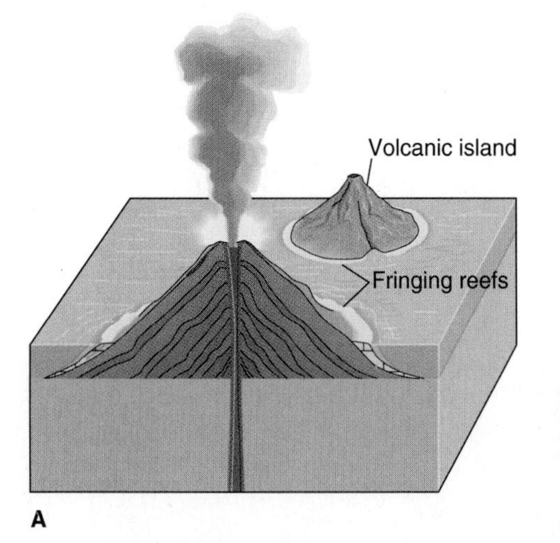

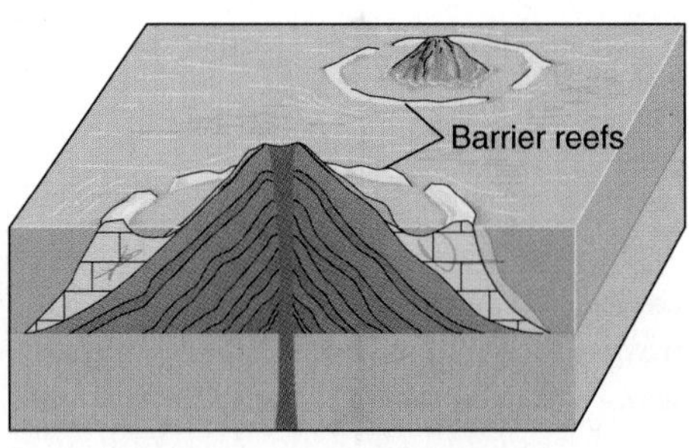

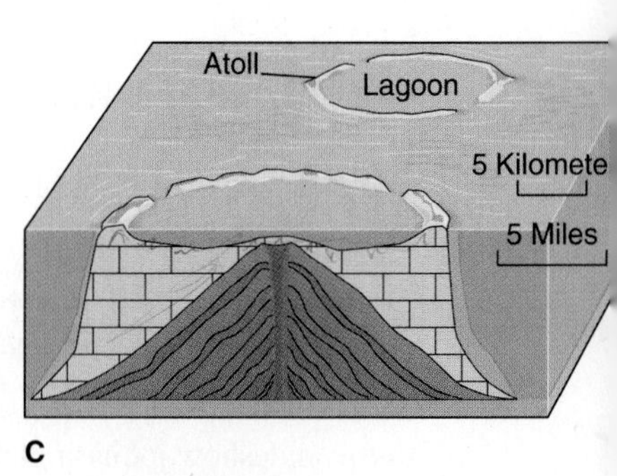

Figure 5.13

Types of coral-algal reefs. (*A*) Fringing reefs are attached directly to the island. (*B*) Barrier reefs are separated from the island by a lagoon. (*C*) Atolls are circular reefs with central lagoons. Charles Darwin proposed that the sequence of fringing, barrier, and atoll reefs form by the progressive subsidence of a central volcano, accompanied by the rapid upward growth of corals and algae.

(From Charles C. Plummer, David McGeary, and Diane Carlson, Physical Geology, 8th ed. Copyright © 1999 McGraw-Hill Company, Inc., Dubuque, Iowa. All rights reserved. Reprinted by permission.)

GLOSSARY

A

Aa. A lava flow, usually basaltic in composition, that has a jagged, rubbly surface. 181

Absolute chronology (absolute ages). Sequential order of geological events (a series of dates, usually in millions of years), based on radiometric measurements. 55, 250

Accessory mineral. A mineral not important in the classification of a rock but may be used as a descriptor. 27

Active dune. A sand dune that migrates in a downwind direction or is constantly changing in shape in response to multiple wind directions at different times. 155

Alluvial fan. A fan-shaped depositional feature produced where the gradient of a stream changes from steep to shallow as the stream emerges from a mountainous terrain. 99

Alpine glacier. A river of ice confined to a mountainous valley. 133

Amygdaloidal. An adjective describing an igneous rock containing amygdules. 27

Amygdule. A vesicle in an igneous rock; filled with secondary mineral matter by groundwater. 27

Angular unconformity. An unconformity in which younger sediments rest upon the eroded surface of tilted or folded older rocks. 56

Annual snow line. The line that separates the accumulation and wastage zone of a glacier for a given budget year. 133

Anticline. A fold, generally convex upward, whose core contains the stratigraphically older rocks. 191

Aphanitic. The texture of an igneous rock composed of microscopic crystals. 27

Arête. A sharp, narrow, rugged ridge (divide) between two parallel glacial valleys. 134

Assimilation. The process in which a magma melts country rock and assimilates the newly molten material. 25

Asthenosphere. The weak, ductile layer of the earth's upper mantle (beneath the lithosphere); the layer on which the lithospheric plates move. 241

Atoll. A roughly circular narrow strip of continuous or broken coral reef with low relief; encloses a lagoon and is bounded on the outside by the deep water of the open ocean. 259

Aureole. The zone of contact metamorphism in the host rock surrounding an intrusion. 45

Avulsion. The process where a river suddenly leaves its old bed and forms a new bed. 100

Axial plane. An imaginary plane that divides a fold into halves as symmetrically as possible. 191

B

Barchan. A crescent-shaped sand dune in plan view with the highest part in the center and the tips (horns) pointing downwind. 154

Barrier island. A long narrow island of sand parallel to the shoreline and built by wave action. 163

Barrier reef. A reef separated from the shoreline by a lagoon. 258

Base level. The lowest level to which a stream can erode its bed; usually sea-level. 98

Batholith. A large mass of intrusive igneous rock that crops out over an area of more than 100 square kilometers. 25

Beach. The wave-washed sediment along a coast, extending throughout the surf zone to a cliff or zone of permanent vegetation. 163

Beach drift. The movement of particles obliquely up the slope of a beach; caused by the swash, and directly down this slope, by the backwash. 163

Bedding plane. A horizontal surface along which a sedimentary rock breaks or separates; represents changes in the depositional history. 35

Bench mark. A relatively precisely located point marked by a brass plate with a cross, elevation, and coordinates permanently fixed on the ground. 75

Birdfoot delta. A delta with many distributaries. 164

Block diagram. A perspective drawing in which the information on a geologic map and geologic cross section are combined. 191

Blowout. A shallow depression on the land surface caused by the removal of sand or smaller particles by wind erosion. 154

Bowen's reaction series. A schematic description of the order in which minerals crystallize and react during the cooling and progressive crystallization of a magma. 24

Breccia. A rock composed of angular fragments of broken rock cemented together. 39

Budget year of a glacier. A 12-month period of winter snow accumulation and summer melting. 133

C

Calcareous. An adjective used for a rock containing calcium carbonate; usually refers to a sedimentary rock in which a carbonate mineral is present but is not a major constituent. 39

Calving. A process whereby masses of glacier ice become detached from a glacier and terminate in deep water to form icebergs. 133

Cirque. A steep-sided amphitheater-like feature carved into a mountain at the head of a glacial valley and formed mainly by glacial abrasion and plucking and by frost wedging. 134

Clay mineral. A group of platy hydrous aluminosilicate minerals that have layered atomic structures that formed from the weathering and hydration of other silicate minerals. 36

Clay size. A particle four microns or less in diameter. 36

Cleavage. (1) The tendency of a mineral to break along certain preferred crystallographic planes. (2) The tendency of a rock to separate into platelike fragments along certain planes; usually the result of preferred orientation of the minerals in the rock. 3; 47

Color. The quality of a mineral or rock with respect to light reflected by it usually determined visually by measurement of hue, saturation, and brightness of the reflected light. 2

Confining (static) pressure. The pressure (equal in all directions) on deeply buried rocks resulting from the weight of the overlying rocks. 45

Confluence. The point at which a tributary joins the main stream. 111

Conglomerate. A coarse-grained rock composed of rounded to subangular clasts set in a fine-grained matrix; the lithified equivalent of gravel. 35

Contact (thermal) metamorphism. The metamorphism of the country rock adjacent to an intrusion of igneous rock; metamorphism in which temperature is the dominant factor. 45

Continental crust. That portion of the crust (outer lithosphere) which is mainly granitic in composition, varies from 20–70 km. thick, and which underlies the continents. 241

Continental glacier. An ice sheet that covers an area of continental proportions and is not confined to a single valley and spreads outwards in all directions under the influence of its own weight. 133

Contour interval. The difference in elevation of any two adjacent contour lines. 71

Contour line. A line on a topographic map connecting points of equal elevation. 71

Core. The central part of the earth below about 2900 km. Composed mainly of iron and nickel. The outer core is liquid, the inner core solid. 241

Country rock. The rock mass intruded by a magma. 25

Crevasse. A deep, gaping, vertical fissure in the upper 40 to 50 meters of a glacier. 141

Cross-bedding. Sediments laid down at an angle to the main sedimentary layering. 37

Crust. The outermost layer of the solid earth (outer lithosphere). 241

Crystal form. In a definite geometric relationship the assemblage of faces that constitute the exterior surface of a crystal. 5

Crystal habit. The crystal form commonly taken by a given mineral. 5

Crystal symmetry. The geometric relationship between crystal faces. 5

D

Deflation. The picking up and removal by wind of loose, dry, surface material. 154

Delta. A body of sediment deposited by a river where it enters an ocean or lake. 164

Detritus. Fragments of minerals, rocks, and skeletal remains of dead organisms. 35

Diaphaneity. The ability of a thin slice of a mineral to transmit light. 5

Dike. A tabular igneous intrusion whose contacts cut across the trend of the country rock. 26

Dip. The vertical angle in degrees measured downward between a horizontal plane and an inclined plane perpendicular to the strike. 190

Disconformity. An unconformity in which the bedding planes above and below a break (visible and irregular or uneven erosion surface of appreciable relief) are essentially parallel, indicating a considerable interval of erosion or nondeposition. 56

Dormant volcano. An active volcano during periods of quiescence. 179

Double refraction. The splitting of light into two components when it passes through certain crystalline substances. 7

Drainage system. A stream together with its tributaries. 98

Drowned river mouth. A long embayment formed by the encroachment of the sea into the mouth of a river. 164

Drumlin. An elongate and streamlined hill of till whose long axis is parallel to the direction of ice movement. 145

E

Effluent stream. A stream fed by groundwater because its channel lies below the water table. 117

End moraine. The accumulation of till at the terminus of a glacier. 133

Epicenter. The point on the earth's surface that lies directly above the focus (hypocenter) of an earthquake. 227

Esker. An ice-contact deposit in the form of a continuous low-ridge of stratified sand and gravel, often sinuous, formed beneath the glacier surface by a sediment-laden stream flowing through an ice tunnel. 145

Essential mineral. A mineral necessary to classify a rock. 27

Eustatic change. A change in sea-level that affects the whole earth. 164

Extinct volcano. A volcano in which all volcanic activity has ceased permanently. 179

F

False Color image. An image from remote sensors combined with other wavelengths in the processing of the imagery data resulting in a colored image in which the true colors are replaced by other colors. 63

Fault. A fracture or break in the earth's crust along which differential movement of the rock masses has occurred. 189

Fault plane. The plane that best approximates the fracture surface of a fault. 189

Fault scarp. The surface expression of the fault plane. 220

Fault trace. The intersection of the fault plane with the ground surface. 219

Felsic. An adjective used to describe a light-colored igneous rock that is poor in iron and magnesium and contains abundant quartz and feldspars. 25

Floodplain. A flat erosional river valley floor on either side of a stream channel; inundated during floods and contains silt and sand carried out and deposited from the main channel during floods. 99

Flow line. (1) The path a water molecule follows from the time it enters the zone of saturation until it reaches a lake or stream where it becomes surface water. (2) The path a particle takes from the point where it is incorporated into a glacier until it ceases movement. 117; 133

Focus (hypocenter). The place where rupture occurs (movement begins) on a fault plane during an earthquake. 227

Foliation. A metamorphic texture in which the mineral constituents are oriented in a parallel or subparallel arrangement. 47

Footwall. The surface of the block of rock below an inclined fault. 219

Fracture. (1) Breakage that forms a surface with no relationship to the internal structure of a mineral. (2) A crack or joint in bedrock. 4

Fringing reef. A coral reef attached to or bordering a landmass. 258

G

Geologic column. A composite diagram that shows in a single column the subdivisions of part or all of geologic time or the sequence of stratigraphic units of a given locality or region (the oldest at the bottom and the youngest at the top) so arranged as to indicate their relations to the subdivisions of geologic time and their relative positions to each other. 55

Geologic cross section. A diagram or drawing that shows geologic features transected by a given plane. 56

Geomorphology. The association of geologic agents with the origin of various landforms. 97

Geothermal gradient. The rate of increase of temperature with depth within the earth. 24

Glacier. A mass of flowing land ice (derived from snowfall) which moves because of its own weight. 133

Graben. A wedge-shaped downthrown block of rock bounded on its sides by normal faults. 220

Graded bedding. Gradual vertical shift from coarse to fine clastic material in the same bed; resulting from deposition by a waning current. 37

Gradient. The vertical drop of a stream over a given horizontal distance. 98

Groundmass. The fine-grained matrix of an igneous rock. 27

H

Hand specimen. A specimen of mineral or rock that can be held in the hand for study. 5

Hanging valley. A valley formed where a tributary glacier once joined the main glacier, and a waterfall cascades from it to the floor of the main valley. 134

Hanging wall. The surface of the block of rock above an inclined fault. 219

Hardness. Resistance of the surface of a mineral to abrasion. 2

Headland. Part of the coast that juts out into a lake or ocean. 163

Heavy minerals. Minerals having a specific gravity greater than 2.85, and commonly found as minor constituents or accessory minerals of a rock. 39

Horn. (1) A sharp pyramid-shaped mountain peak near the heads of valley glaciers or glaciated valleys. (2) The ends of a barchan. 134

Hornfels. A fine-grained nonfoliated rock usually formed by contact metamorphism. 46

Horst. An upthrown block of rock bounded on its sides by normal faults. 220

Hot spot. A deep-seated zone of intense heat whose geographic coordinates remain fixed for several million years. 179

I

Iceberg. A mass of glacier ice calved into a lake or ocean. 133

Igneous rock. A rock formed by the cooling and crystallization of molten material within or at the surface of the earth. 23

Inactive dune. A sand dune that has become stabilized by the growth of vegetational cover to the extent that its migration ceases. 155

Inactive fault. A fault in which no movement has occurred during historical times. 219

Inclusion (xenolith). A fragment of country rock surrounded by igneous rock. 25

Index mineral. A mineral that forms or is stable over a limited range of temperature and pressure conditions and whose first appearance marks the outer limits of a specific zone of metamorphism. 46

Influent stream. A stream that lies above the water table and in which groundwater flows away from the stream channel. 117

Intermittent stream. A stream that flows only during certain times of the year when rainfall is sufficient to supply surface runoff directly to it. 117

K

Kame. A knob, hummock, or conical hill composed of coarse stratified drift formed as a delta at the front of a glacier by meltwater streams. 145

Karst topography. A terrain marked by many sinks and caverns and usually lacking a surface stream. 123

Kettle. In an end moraine a depression formed by the melting of an underlying large block of ice left behind by a receding glacier. 145

L

Laccolith. A lenticular pluton, many times wider than it is thick, intruded parallel to the layering of the intruded rock, above which the layers of intruded rock have been bent upward to form a dome. 26

Lagoon. A shallow water body lying between the main shoreline and a barrier island or inshore from an enclosing reef. 163

Lateral moraine. An accumulation of till along the sides of an Alpine glacier. 133

Lava fountain. Lava spouting from a fissure in a rift zone as a result of gas and fluid pressure that builds up in the crust below. 180

Law of constancy of interfacial angles. The angle between similar crystal faces of a mineral is constant. 5

Law of original horizontality. Sediments deposited in water are laid down in strata that are horizontal or nearly horizontal and are parallel or nearly parallel to the Earth's surface. 55

Law of superposition. In any undisturbed sequence of sedimentary rocks, the layer at the bottom of the sequence is older than the layer at the top of the sequence. 55

Leeward. The downwind side of a sand dune. 154

Lithification. The combination of processes that convert a sediment into a sedimentary rock. 23

Lithosphere. The rigid outer 100 km of the solid earth consisting of the rigid upper mantle and crust of the earth. 241

Longitudinal (seif) dune. A long linear symmetrical dune with its long axis parallel to the wind direction. 155

Longitudinal profile. A line showing a stream's slope, drawn along the length of the stream as if it were viewed from the side. 98

Luster. The appearance of a fresh mineral surface in reflected light. 2

M

Macroscopic (megascopic). A feature of a mineral or rock that can be distinguished without the aid of magnification. 1

Mafic. An adjective used to describe a dark-colored silica-poor igneous rock with a high magnesium and iron content and composed chiefly of iron- and magnesium-rich minerals. 25

Magma. Molten material generated within the earth and consisting of a complex solution of silicates plus water, dissolved gases, and any suspended crystals. 24

Magnetic anomaly. A magnetic reading that is greater or less than the normal strength of the magnetic field where the rock occurs. 250

Magnetometer. An instrument that measures the intensity of the earth's magnetic field. 250

Mantle. The largest portion of the solid earth, separating crust above from the core below. It is mainly ultramafic in composition. 241

Meander. A looplike bend of a stream channel; develops as the stream erodes the outer bank of a curve and deposits sediment against the inner bank. 100

Meandering course. The circuitous course of a stream flowing across its flood plain. 100

Medial moraine. A linear moraine formed by the joining of two lateral moraines as the two glaciers flow together. 133

Mesosphere. The portion of the mantle between the bottom of the asthenosphere and the core-mantle boundary. 243

Metamorphic facies. The pressure and temperature stability fields for metamorphic rocks as determined by mineral assemblages. 45

Metamorphic grade. The intensity of metamorphism in a given rock; the maximum temperature and pressure attained during metamorphism. 45

Metamorphic rock. A rock resulting from the change of a preexisting rock as a result of the effects of heat, pressure, chemical action, or combinations of these. 23

Metasomatic rock. A rock whose chemical composition has been substantially changed by the metasomatic alteration of its original constituents. 45

Metasomatism. The process of solution and deposition by which a new mineral or minerals of partly or wholly different chemical composition may grow from externally supplied fluids and elements in the body of an old mineral or mineral aggregate. 45

Microscopic. A feature of a mineral or rock that can be distinguished only with the aid of magnification. 1

Mohs scale of hardness. Ten common minerals arranged in order of their increasing hardness. 2

Monocline. A one-limb flexure in which the strata have a uniform direction of strike but a variable angle of dip. 191

N

Natural levee. A narrow ridge of flood-deposited sediment found on either side of a stream channel, which thins away from the channel. 164

Nonconformity. An unconformity developed between sedimentary rocks and older plutonic igneous or metamorphic rocks that had been exposed to erosion before the overlying sediments covered them. 56

Nonfoliated. No preferred orientation of mineral grains in a metamorphic rock. 49

Normal fault. A fault in which the hanging wall has moved downward relative to the footwall. 220

O

Oceanic crust. The crust (outer lithosphere) beneath the oceans consisting of a 7–10 kilometer thick layer of mainly basalts and thin sedimentary rocks and overlying sediments. 241

Oolitic. A sedimentary texture formed by spheroidal particles (oolites) less than 2 mm in diameter. 36

Outwash. A deposit of sand and gravel formed beyond the front of a glacier by streams flowing from the glacier terminus. 145

Oxbow lake. A crescent-shaped shallow lake formed where a meander loop is cut-off from the main stream and its ends plugged with sediments. 100

P

Pahoehoe. A smooth surface "ropy" lava usually basaltic in composition. 181

Paleomagnetism. The study of natural remanent magnetism in ancient rocks, recording the direction of the magnetic poles at some time in the past. 250

Pediment. A gently sloping erosional surface cut into the solid rock of a mountain range and covered with a thin, patchy veneer of coarse colluvium that slopes away from the base of a highland in an arid or semiarid environment. 99

Pegmatite. A very coarse-grained igneous rock of any composition but most frequently granitic. 29

Phaneritic. The texture of an igneous rock composed of macroscopic minerals. 27

Plan view. As viewed from above. 81

Plate tectonics. A theory that the earth's surface is divided into a few large, thick, rigid plates that are slowly moving and changing size with respect to one another. 241

Plunging fold. A fold with an inclined axis. 191

Pluton. A body of igneous rock, regardless of shape or size, that crystallized underground. 25

Porphyritic. An igneous rock in which phenocrysts (macroscopic minerals) are embedded in a fine-grained matrix. 27

Porphyroblast. A metamorphic mineral crystal that is much larger than the matrix in which it grew. 49

Primary minerals. Minerals that crystallized from a cooling magma. 27

Prograding shoreline. A shoreline that moves seaward by deltaic growth. 165

Pyroclastic. Accumulations of material ejected from explosive-type volcanoes. 29

R

Recessional moraine. Successive end moraines lying beyond the snout of a retreating glacier. 133

Recumbent fold. An overturned fold in which the axial plane and both limbs are essentially horizontal. 191

Relative chronology (relative ages). Sequential order of geological events based on relative position, fossil content, and cross-cutting relationships. 57

Remote sensing. A process whereby the image of a feature is recorded by a camera or other devices that are not in direct contact with it and reproduced in one form or another as a "picture" of the feature. 63

Reverse fault. A fault in which the hanging wall has moved up relative to the footwall. 220

Ripple marks. The parallel ridges and troughs left by running water or wind on a former sedimentary surface. 37

Rock cycle. The sequence of events in which rocks are created, destroyed, altered, and reformed by geological processes. 23

S

Sand dune. A ridge or mound of sand deposited by wind. 154

Secondary mineral. A mineral formed later than the rock enclosing it, usually at the expense of an earlier formed primary mineral. 27

Section. A rectangular block of land one mile long by one mile wide. 68

Sedimentary rock. A rock formed from precipitation from a solution, by accumulation of organic materials, or by sedimentation and cementation of sediments derived from preexisting rocks and transported to a site of deposition by water, wind, or ice. 23

Serrate divide. The divide between the headward regions of oppositely sloping glacial valleys. 134

Shield volcano. A broad gently sloping dome built of thousands of highly fluid (low viscosity) lava flows generally basaltic in composition. 179

Sialic rock. A light-colored igneous rock that contains large proportions of silicon and aluminum. 25

Sill. A tabular shaped igneous rock mass that lies parallel to the layering of the intruded rock. 26

sink (sink hole). A surface depression in limestone caused by the collapse of the roof of an underground dissolution cavity. 123

Slip face. The steep leeward slope of a sand dune formed from loose, cascading sand that generally keeps the slope at the angle of repose (about 34°). 154

Sole marks. Marks left at the top of a soft sediment by bottom-dwelling organisms. 37

Specific gravity. A number stating the ratio of a mineral's weight to the weight of an equal volume of pure water. 4

Spit. An elongate ridge of sand or gravel extending from shore into a body of open water; deposited by longshore currents. 163

Stack. An isolated rocky pillar that is the remnant of a retreating wave-cut cliff. 164

Stock. An irregular body of intrusive igneous rock, smaller than a batholith, that cuts across the layering of the intruded rock. 25

Stratovolcano (composite volcano). A conical mountain with steep sides composed of interbedded layers of viscous lava and pyroclastic material. 179

Streak. The color of a mineral's powder usually obtained by rubbing the mineral on a streak plate (unglazed porcelain). 4

Strike. A horizontal line in a plane of the bedding or fault surface expressed as a compass direction from true North. 190

Strike-slip fault. A fault with relative displacement along the fault plane in a horizontal direction parallel to the strike of the fault plane. 220

Subduction zone. A zone where oceanic lithosphere descends beneath a continental or island arc along a deep ocean trench; defined by high seismicity. 241

Swash. The surge of water up a beach caused by waves breaking against a coast. 163

Syncline. A fold, generally concave upward, whose core contains the stratigraphically younger rocks. 191

T

Tarn. A small lake within a cirque. 134

Tenacity. A mineral's resistance to being broken or bent. 5

Tephra. A collective term assigned to all sizes of airborne volcanic ejecta. 181

Texture. The appearance of a rock; results from the size, shape, and arrangement of the mineral grains or crystals. 27

Thalweg. A line connecting the deepest parts of a stream. 111

Thrust fault. A reverse fault in which the fault plane dips less than 45 degrees over most of its extent. 220

Tidal inlet. A break or passageway through a bar or barrier island; allows water to flow alternatively landward with rising tide and seaward with falling tide. 163

Till. The nonsorted debris carried and deposited directly by a glacier. 133

Topographic map. A map on which elevations are shown by means of contour lines. 70

Topography. The relief and form of the land surface. 70

Township. An area containing 36 sections. 68

Transform fault. A strike-slip fault that offsets spreading ridges. 220

Transverse dune. A linear ridge of sand with a gently sloping windward side and a steep lee face with its long dimension oriented perpendicular to the prevailing wind direction. 155

Tuff. The consolidated ash and pyroclastic fragments generated from a volcanic eruption. 29

U

Unconformity. A surface that represents a break in the geologic record, such as an interruption of deposition of sediments or a break between eroded igneous and overlying sedimentary rocks. 56

U-shaped valley. A glacially eroded valley with a characteristic U-shaped cross section. 134

V

Vesicle. A small cavity in an igneous rock, usually extrusive, that was formerly occupied by a bubble of escaping gas originally held in solution under high pressure while the parent magma was deep underground. 27

Vesicular. A texture characterized by the presence of vesicles. 27

W

Wastage zone. The area of a glacier that suffers wastage of both snow and ice. 133

Water table. The top of the zone of ground water saturation. 117

Wave-cut cliff. A cliff formed by wave refraction against headlands. 163

Wave-cut platform. A gently sloping rock surface lying below sea level and extending seaward from the base of a wave-cut cliff. 164

Windward. The side of a dune ridge facing the wind. 154

Z

Zone of saturation. Beneath the surface of the earth, the zone that is saturated with water. 117